D1704704

GEOMORPHIC PROCESSES II

Adrian M. Harvey & Maria Sala (Editors)

GEOMORPHIC PROCESSES

In Environments With Strong Seasonal Contrasts

Vol. II: GEOMORPHIC SYSTEMS

Selected papers of the "Regional Conference on Mediterranean Countries", IGU Commission on Measurement, Theory and Application in Geomorphology, Barcelona–Valencia–Murcia–Granada, September 5–14, 1986

CATENA SUPPLEMENT 13

CATENA – A cooperating Journal of the International Society of Soil Science

ISSS - AISS - IBG

Cover photo by Heinrich Rohdenburg:
Slope denudation; accumulation and incision of torrents of different hierarchies, near Almeria, Southeast Spain

CIP-Titelaufnahme der Deutschen Bibliothek

Geomorphic processes in environments with strong seasonal contrasts: selected papers of the "Regional Conference on Mediterranean Countries", Barcelona, September 1986. -
Cremlingen-Destedt: Catena.
NE: Regional Conference on Mediterranean Countries <1987, Barcelona>
Vol. II. Geomorphic systems / Adrian M. Harvey & Maria Sala (ed.). - 1988
(Catena: Supplement; 13)
ISBN 3-923381-13-1
NE: Harvey, Adrian M. [Hrsg.]; Catena / Supplement

ISSN 0722-0723 / ISBN 3-923381-13-1

CONTENTS

Preface

In September 1986, the IGU Commission on Measurement, Theory and Application in Geomorphology (COMTAG) organised its main autumn meeting in Spain. This meeting had as its theme "Geomorphological processes in environments with strong seasonal contrasts". Most of the papers presented at the meeting are published in two Special CATENA SUPPLEMENTS, of which this is the second. Volume I edited by A.C. IMESON and M. SALA contains papers dealing primarily with hillslope processes **per se**. This volume contains papers dealing with sediment transport through fluvial systems and the relationships between processes and morphological development in fluvial systems. These, of necessity have a broader spatial or temporal context than those in volume I.

The COMTAG meeting in Spain was organised principally in Barcelona, Murcia and Granada where paper sessions were held. A major feauture of the meeting was the large number of field visits to sites where investigations were in progress. In this way the COMTAG symposium drew attention to the interesting and important work being done in Spain and at the same time served to stimulate research in process geomorphology. During the meeting the seminal contribution made by Professor Luna B. LEOPOLD to our understanding of geomorphic processes, especially in the drier regions of the world, was formally recognised by the University of Murcia. At a ceremony in Murcia he was created Honorary Professor of that University. The Spanish text of his address is presented elsewhere (LEOPOLD 1986). At this point the organisers of the COMTAG meeting in Spain and the editors of this volume wish to thank Professor Leopold for his guidance, leadership and inspiration in the field of process geomorphology.

The location of the meeting in Spain provided the opportunity for COMTAG to focus its attention on regions with strong seasonal contrasts. Geomorphic systems in such environments are highly complex and difficult to study due to seasonally extreme conditions. The meeting reflected the need to increase our understanding of the basic processes in these regions, of how seasonally variable processes may interact and of how such interactions may influence landform development.

The eleven papers which comprise this volume have been arranged in relation to spatial and temporal scales. The first four papers all deal with erosion and sediment transport in badland catchments. ROMERO-DIAZ et al. and BRYAN et al. consider relatively short-term erosion rates and processes in southeast Spain and in Canada and Kenya respectively. CLOTET-PERERNAU et al. working in the Spanish Pyrennees, compare short and medium-term timescales and GUTIERREZ et al. consider the influence of piping on the morphological development of badlands in the Ebro Basin of northern Spain. The next two papers deal with geomorphic response to meteorological sequences, SUWA & OKUDA with process-seasonality in Japan and GALLART & CLOTET-PERERNAU with response to an extreme rain in the Spanish Pyrennees. The following three papers all consider the relationships between sediment source areas and depositional sites. ERGENZINGER specifically considers temporal and spatial scales in sediment production, transport and erosion rats in Calabria, SORRISO-

VALVO, also working in Calabria and HARVEY working in southeast Spain both deal with alluvial fans. Next SANCHO et al. consider long-term slope development in the Ebro basin and finally CONACHER examines the significance of short-term process measurements to the understanding long-term landform development in Western Australia.

Several papers presented at the meeting are not included in this volume and are being developed for publication elsewhere. These include papers by GUPTA on channel morphology in India, SEGURA on stepped channels in Spain and BORDAS on Thermo-Luminescence dating of soils in Spain.

References

LEOPOLD, L.B. (1986): Algunes observaciones sobre los paisajes semiarides. Papeles de Geografia Fisica, Murcia, **11**, 5–6.

Adrian M. Harvey
Dept. of Geography, University of Liverpool
Liverpool, England

Maria Sala
Dept. de Geografia Fisica, Universidad de Barcelona
Barcelona, España

CATENA SUPPLEMENT 13 p. 1–11 Braunschweig 1988

VARIABILITY OF OVERLAND FLOW EROSION RATES IN A SEMI-ARID MEDITERRANEAN ENVIRONMENT UNDER MATORRAL COVER MURCIA, SPAIN

M.A. **Romero-Díaz**, F. **López-Bermúdez**, Murcia
J.B. **Thornes**, C.F. **Francis**, Bristol
G.C. **Fisher**, Egham †

Summary

This paper reports on measurements of erosion at two time and space scales in a semi-arid mediterranean environment. The relationships between rainfall amount and sediment loss are evaluated for the three years data collected to date for a field site near Murcia, south east Spain. The area is an experimental plot 50 m × 60 m developed in marls and covered with sandstone particles which have been broken down from a caprock. Within this eleven small runoff plots with Gerlach troughs have been established with a variety of slope, vegetation cover and roughness conditions. The plot drains to a set of two tanks with capacities of 83.4 and 87.5 litres and four 200 litre sediment and water collectors. Rainfall volumes and intensities were measured at the site. The field monitoring commenced in January 1984 and was discontinued in January 1987. The results show that at the trough scale, values of sediment yield vary greatly, though systematically, and are poorly related to each other and to rainfall. At catchment scale and for individual events the relationship between rainfall and sediment yield is stronger. When the data at this scale are integrated to monthly periods the correlation between rainfall totals and sediment yields are good.

ISSN 0722-0723
ISBN 3-923381-13-1

D–3302 Cremlingen-Destedt, W. Germany
3-923381-13-1/88/5011851/US$ 2.00 + 0.25

Resumen

El trabajo que se presenta hace referencia a mediciones de la erosión del suelo llevadas a cabo a dos escalas temporales y espaciales en un medio mediterráneo semi-árido. Se evalúan las relaciones entre la cantidad de precipitación y la pérdida de sedimento para los tres años en que se han recogido datos en una estación situado cerca de Murcia, en el sureste de España. Esta área consiste en una parcela experimental de 50 m × 60 m emplazada sobre un sustrato de margas cubierto por derrubios de arenisca procedentes de la rotura de un saliente rocoso. Se han instalado once cajas Gerlach situadas en puntos con diferencias en vegetación, pendiente y rugosidad del terreno. Cuatro colectores de 200 litros y dos de 83,4 y 87,5 litros situados al pie de la ladera recogen el agua y sedimento procedente de la parcela. También se tomaron mediciones en el mismo lugar de los volúmenes e intensidades de la precipitación. El control de campo se inició en 1984 y se continúa en la actualidad. Los resultados muestran que, a escala de los canales, los valores de producción de sedimento varían mucho, aunque de manera sistemática, y las relaciones entre ellos y con la precipitación son bajas. La relación es mayor cuando se mira a escala del conjunto de la parcela e individualmente para cada precipitación por separado. Cuando los datos a esta escala global se integran en periodos mensuales la correlación entre totales de precipitación y producción de sedimento mejora sensiblemente.

1 Introduction

Mediterranean environments are characterised by strong seasonal contrasts in climate. When this is coupled with low annual rainfall and rocks of high erosional susceptibility the potential for erosion is very high. In south-east Spain, which has these characteristics in combination and where there has been a long history of human effects on the vegetation cover and soils, it is widely accepted that there has been continued and more or less intensive erosion since Neolithic times. However there have been several suggestions that here, as in other apparently rapidly eroding areas of the world, the rates of erosion are highly variable in both time and space and that an unvegetated and heavily gullied landscape is not necessarily indicative of high contemporary rates of erosion (YAIR, GOLDBERG & BRIMMER 1982). Until more data are available from detailed studies of reservoir sedimentation, well documented historical studies and detailed investigations of contemporary processes and water and sediment production, speculative arguments are likely to continue about, for example, the relative merits of different methods of combatting erosion.

The relatively scarce measurements of contemporary rates of erosion in semi-arid mediterranean Spain come mainly from estimates of reservoir sedimentation made by the Centro de Estudios Hidrográficas y Confederaciónes Hidrograficas (LÓPEZ-BERMÚDEZ & GUTIERREZ-ESCUDERO 1982).

Other estimates have been obtained from the sediment yields of rivers and from a variety of formulae, including the U.S.L.E. (ICONA 1982, GRUPO DE TRABAJO REGIONAL DEL SEGURA 1985, LÓPEZ-BERMÚDEZ 1986). Finally, and only recently, field experiments have begun to contribute data on contemporary rates. These various sources of estimates are given in tab.1. In this paper we provide the results of a set of measurements derived from our own field experiments in the Province of Murcia in the period 1984–1987. In a semi-arid environment this is a very short record but the results tend to confirm the trends observed in other semi-arid areas and throw some light on the recent debate relating to the relationship between observation made at different spatial and temporal scales (ROELS 1985). For these reasons the results may be of more than purely local interest.

Author	Area	Method	Lithology	Rate ($Tm\ ha^{-1}\ yr^{-1}$)
FRANCIS (1986)	Murcia	Gerlach Troughs	Marl	1.8–3.2 (for 1985/86)
JANSEN & PAINTER (1974)	Various		Various	0.25
LÓPEZ-BERMÚDEZ (1987)	Segura Basin (Murcia)	Reservoir Sedimentation	Various	2.0-14.0
SCOGING (1982)	Ugijar (Granada)	Erosion pins	Various	36.9–94.6
WISE et al. (1982)	Posito River (Granada)	Sediment Load	Various	0.16–0.40

Tab. 1: *Rates of erosion for various lithologies and techniques.*

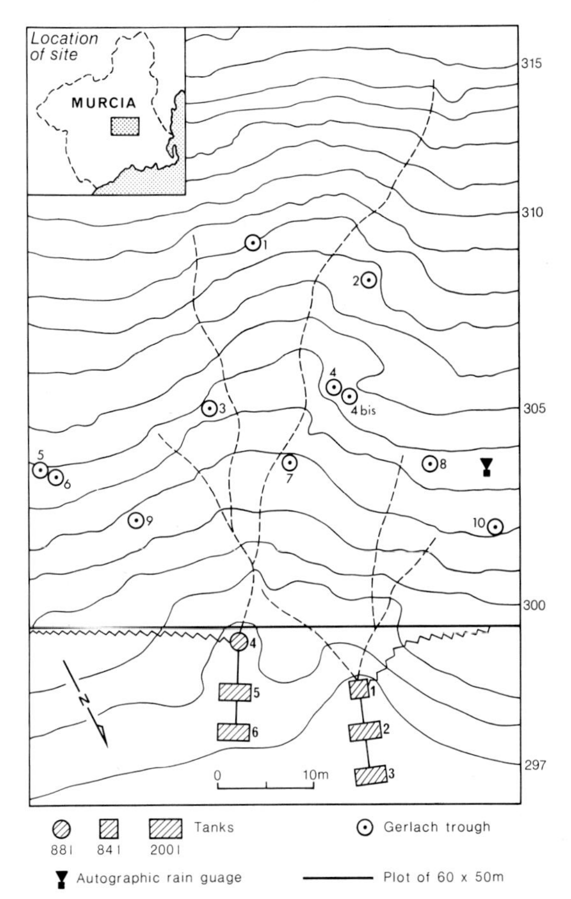

Fig. 1: *Plan of site indicating the positions of the Gerlach troughs and the collectors: inset shows the general location.*

Tanks are 88, 84 and 200 litres.

2 Environmental Conditions

The field site is situated in the eastern part of the Mula basin, in the Province of Murcia (fig.1). The mean annual rainfall is just under 300 mm, the mean annual temperature 18°C, and the potential evaporation about 900 mm y^{-1}, the rainfall being seasonally distributed with the period from late April to September being dominantly rain free. This long and excessively dry priod is occasionally punctuated by heavy and intense convective rainfalls in late August or early September which cause heavy flooding and high levels of damage. The rainfall is also highly variable from year to year (25 to 30%) according to the long record evaluated by LÓPEZ-BERMÚDEZ (1971). The experimental site comprises a shallow basin set into a cuesta formed of Neogene marine marls overlain by a sandstone caprock. The latter sheds coarse particles onto the footslopes into which the small catchment is eroded. Slopes are typically about 15°. The amount of vegetation cover is about 35%, fluctuating during the year as a function of the summer drought and consists of a low and discontinuous matorral. The dominant species are *Brachypodium* sp. and *Thymus* sp. Other species present include *Stipa tenacissima, Fumana thymifolia, Sideritis leucantha*, and *Salsola genistoides*. The soils are entisols with little differentiation between horizons despite appreciable quantities of organic matter in the upper 25 cm. The underlying rock has a high clay content and a mean calcium content of 55%, and the soils are alkaline in reaction. Further details of the site and its vegetation are to be found in LÓPEZ-BERMÚDEZ et al. (1985) and FISHER et al. (1987), and of the soil-moisture, litter and organic matter relationships in FRANCIS et al. (1986). Essentially the stony site, its topography, heavily-grazed vegetation cover and erosional-ecological relationships are typical of large areas of semiarid mediterranean Spain.

3 Methods

The rainfall for the first twenty months of the experiment was determined by inspection of an ordinary rain collector as soon as possible after each event. For this reason the intensities had to be estimated from nearby rainfall stations. However in October 1985 an autographic recorder was installed and intensities could thereafter be obtained straight from the chart.

The experimental site is a rectangular plot 60 m × 50 m with an average slope of 15°. The plot is limited laterally by two small divides and at the top and bottom by small man-made ridges which prevent water flowing onto the plot from upslope and to channel the runoff and sediment produced within the plot to collecting points at the bottom of the slope (fig.1). The study was initiated at the beginning of 1984 with the construction of two water collectors with capacities of 83.4 and 87.5 litres at the foot of the slope. Subsequent observations showed that these were not sufficiently large and a further four collectors were added to bring the total to six, with two groups of three collectors placed in two small runoff channels. The four additional containers each have a capacity of 200 litres bringing the total available storage for water and sediment to 972 litres. For 20 out of 37 storms this capacity was exceeded. In these cases the runoff and sediment production will be underestimated. This underestimation is likely to be very

Trough No	Rank 1985	Rank 1986	Correlation Coefficient r	N	Significance Level	Total Sediment 1985–1986 grms
1	5	3	0.55	17	5%	2983.2
2	8	5	0.59	17	5%	1884.6
3	3	7	0.24	17		2539.8
4	9	10	0.31	16		1247.6
4b	7	9	0.30	17		1346.4
5	1	1	0.58	17	5%	8334.1
6	2	2	0.55	17	5%	4105.4
7	11	8	0.62	16	5%	1196.3
8	6	6	0.41	16		1956.5
9	10	11	0.31	17		912.3
10	4	4	0.51	15		3231.3

Tab. 2: *Rank values, statistical data and total soil loss for the Gerlach troughs in 1985 and 1986.*

small for sediment yield, since an analysis of the data for smaller storms shows that most of the sediment is caught by the first two collectors. It is impossible to know what the underestimation is for water and sediment concentration.

Within the site eleven Gerlach troughs were installed in May 1984. The locations of these troughs were established by taking into account interfluve and slope sections, the ground roughness and vegetation cover conditions. For both these and the large collectors, runoff and sediment yield were collected after each storm event and, after removal of the organic matter by flotation, the quantity of sediment in each collector was determined by weight. Coarse particle sizes have been obtained by separation and weighing. For finer particles however, with the high calcium content, there is a tendency for dried fines to aggregate into crusts which can only be broken down mechanically, after which it is impossible to know the true original aggregate sizes. This is a common problem in semi-arid environments.

4 Results

The Gerlach trough results are shown in tab.2. The table shows the relative ranking of the troughs in 1985 and 1986 with respect to sediment receipts, the correlation with rainfall amount through time for each trough and the total sediment caught by each trough over the two years. There are two points to make about these results. First, when all the troughs are aggregated by event, then the correlation with rainfall amount for the linear model is 0.54. This is similar for the plot as a whole as will be seen below. The regression is significant at the 5% level but not very strong. Second, decomposition of the data by regression of rainfall amount and sediment by storm for each trough taken separately shows a wide variety of values of the linear coefficient, indicating the relatively high local noise effects induced probably mainly by variations in ground cover and roughness, especially stoniness. If this is local spatially-dominated noise then the sediment amounts caught in the troughs should be consistent from one year to

Date	Rainfall (mm)	Sediment Yield (kg)	Date	Rainfall (mm)	Sediment Yield (kg)	Date	Rainfall (mm)	Sediment Yield (kg)
1984			**1985**			**1986**		
27 Feb	4.7	0.16	6 Jan	3.0	0.00	30 Jan	9.4	9.31
14 Mar	0.5	0.00	26 Jan	2.3	0.00	4 Feb	0.4	0.00
20 Mar	6.1	0.00	10–11 Feb	3.4	0.00	21 Feb	1.8	0.00
22 Mar	2.4	0.00	20–21 Feb	30.8	4.48	1 Mar	2.0	0.00
28 Mar	3.8	0.02	25 Feb	0.6	0.00	7–9 Mar	36.4	9.44
15 Apr	7.6	0.08	13 Mar	6.0	0.01	20 Mar	5.5	0.00
29–2 May	33.2	6.43	4 May	23.5	177.18	6 Apr	9.5	1.75
3 May	9.9	4.93	10 May	22.0	12.25	29–30 May	55.5	21.91
8 May			23 May	10.2	0.04	7 Jun	5.4	0.00
10 May	2.3	0.00	24 Sep	17.5	111.71	13 July	2.4	0.00
11 May	3.5	7.02	20 Oct	4.3	1.07	26 July	8.5	14.57
19 May	6.2	0.02	26 Oct	27.2	364.76	27 Sep	0.6	0.00
29–30 May	14.0	0.06	4 Nov	0.4	0.00	29 Sep	0.4	0.00
17 Jun	4.7	0.00	14 Nov	28.5	8.06	4–5 Oct	109.6	291.05
18 Jun	1.7	0.00	16 Nov	27.5	12.79	9 Oct	4.5	0.27
21 Oct	3.6	0.01	25 Nov	3.3	0.00	11 Oct	3.0	
31–1 Nov	7.8	2.86	26–27 Nov	3.4	0.00		2.0	
5 Nov	5.7	0.34	15–16 Dec	3.2	0.00		2.2	
9–10 Nov	22.3	2.51	22 Dec	1.0	0.00		103.3	228.34
11 Nov	2.1	0.00	23 Dec	0.3	0.00	15 Oct	5.5	0.51
			29 Dec	19.5	7.07	16 Oct	5.8	21.01
						15 Nov	2.8	0.00
						17 Nov	4.5	0.00
						13 Dec	3.6	0.00

Tab. 3: *Rainfall and sediment data for the experimental site between 1984 and 1986.*

the next. This is born out by a comparison of the Spearman's rank correlation coefficient for the ranking of troughs by sediment catches in the years 1985 and 1986. The value of Rs is 0.8 which is significant at the 0.01 level. This ranking largely reflects the high values of the interfluve troughs where a cover of smaller surface stones and lower amounts of vegetation tend to induce higher rates of erosion. Clearly while in some cases the correlation with rainfall amount is good, in others it is virtually zero. In order to get a more meaningful measure of erosion therefore it is necessary to integrate up to the catchment scale and hence effectively cancel out the local effects.

At the catchment scale the data are obtained from rainfall and sediment caught in the large collectors. The relevant data are given in tab.3 and fig.2. These indicate that in 1984 the sediment production was relatively low, with 0.08 t ha yr^{-1}, whereas in 1985 and 1986 it was 2.55 and 2.36 respectively. These figures are the same order of magnitude as those obtained from reported reservoir sedimentation in the Segura Basin by LÓPEZ-BERMÚDEZ (1987) and by FRANCIS (1986) for abandoned fields. These may simply reflect the annual rainfall totals. The year 1984 was rather dry, the study area receiving only 142.1 mm compared with a normal year of about 300 mm. By contrast 1985 and 1986 had much higher values with 288.4 and 445.4 mm respec-

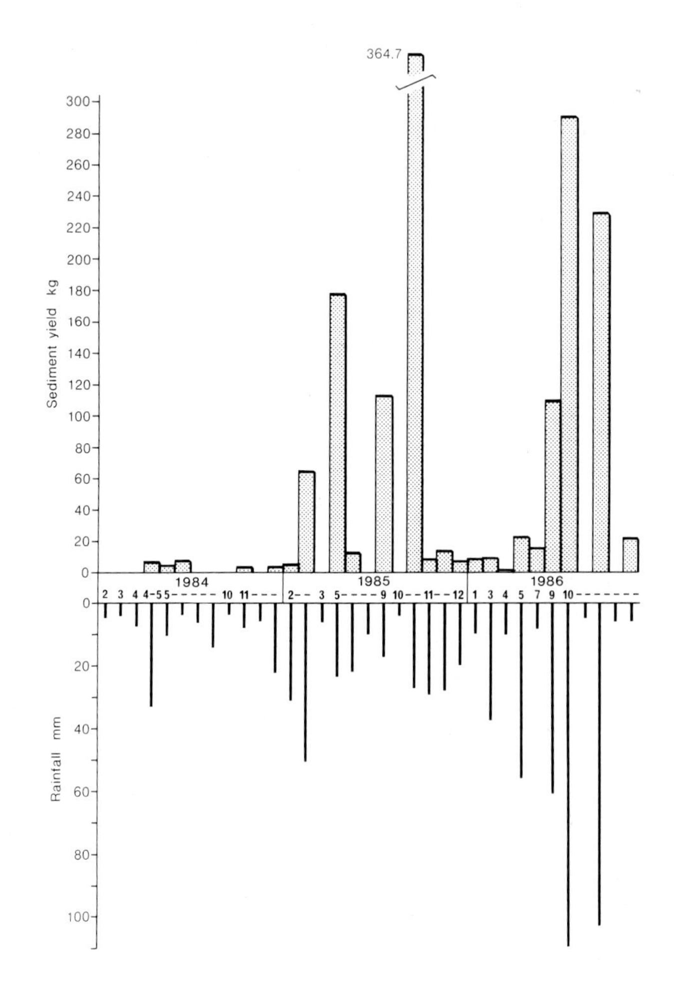

Fig. 2: *Diagram illustrating individual runoff events and sediment yield produced by them, excluding those events which produced no sediment yield.*

NB the time scale on this diagram is nominal.

tively. These broad figures tend to obscure a rather significant interstorm variability. Regression analysis of amount of sediment yield against rainfall on a storm basis reveals a correlation coefficient of 0.65, which is significant at the 99.9% level. The form of the regression model indicates that no significant erosion takes place with rainfalls of less than about 4 mm. Again the relationship is best fitted by a linear model. The correlation of sediment yield with intensity gives a value of 0.56, but one must bear in mind that for the first part of the data intensities are estimated from adjacent stations.

The interstorm variability is best illustrated by a comparison of two storms. In 1984 a small storm of 3.5 mm falling in 10 minutes produced 7.02 kg of sediment, while a storm of 6.2 mm falling over 3h 30m produced only 0.02 kg of sediment. In 1985 and 1986 the storms were greater and sediment production was substantial. For example on 4th May 1985 a storm of 23.5 mm and intensity of 35.2 mm/h eroded a total of 177.18 kg of sediment and secondly on 26th October 1985, 27.2 mm of rain fell with an intensity of 18.13 mm/h eroding 364.76 kg. This storm deposited stones greater than

3 cm in diameter in the first tanks, underlining the high erosive potential of these relatively extreme geomorphic events.

The production of sediment is also a function of the nature of the soil itself, and it is the variability of the conditions of the soil prior to different storms which partially account for the correlation between rainfall and sediment production. The storms which eroded the largest quantities of sediment are those which were precceded by a long period of drought. For example after 123 days of dry weather 17.5 mm of rain fell on 24th September 1985, with an intensity 8.75 mm/hr, which eroded 111.71 kg of soil. This reflects the loss of coherence of the surface soil particles due to mechanical, chemical and biotic disturbance during the previous 123 consecutive days of dry weather. IMESON et al. (1982) and IMESON & VERSTRATEN (1985) have shown the importance of the prior preparation of soil surface conditions for runoff and sediment yield generation in this type of environment. When the analysis is extended however to all the data, the partial coefficient relating sediment yield to the number of previous dry days, with rainfall held constant, is negligibly small. We conclude from this that the role played by antecedent conditions is fairly complex.

5 Rainfall and Soil Loss Budget for the Site

We have shown that the Gerlach trough data are dominated by variability at the local spatial scale. Equally there are substantial sources of variation between subcatchment of the scale of several thousands of square metres. We are therefore rather reluctant to extrapolate our data spatially. However we believe that the data obtained are probably reasonable estimates for this plot in these years despite the low values of sediment yield of about 2.35 t ha yr^{-1}.

The question arises as to whether or not the storms were characteristic of a longer run of years. To test this we have looked at the daily rainfall data for the plot over the years of the experiment and at the nearest detailed station, Murcia, for the past 35 years. The comparative frequency of storms of different magnitudes are shown in tab.4. They indicate that the longer term records had a larger percentage of daily rainfalls (72.9%) of less than 5 mm when compared with the experimental plot with only 50.7%. This was largely made up by a higher frequency of storms in the class 5.00–9.99 mm on the plot. This in turn suggests that since storms less than about 4 mm produce very little sediment the period under investigation may have been more erosive than usual. This is especially true if the two large storms at the very end of 1986 are taken into account. With 109.6 and 103.3 mm they are estimated to have recurrence periods of about 100 years when compared with data for the surrounding stations (ELIAS CASTILLO & RUIZ BELTRAN 1979).

Taken as a whole, these hillslopes, although eroded along the sheep trail at their foot, largely comprise matorral-covered rock-strewn slopes in which runoff is relatively small and where the roughness and resistance to transport is very large. The coefficient of runoff for those storms where the complete data was available are extremely low implying, in addition, relatively high losses to evaporation and infiltration. These slopes are not comparable to the deeply

Clase	% Storm Frequency Experimental Site	% Storm Frequency Murcia
0.0– 4.9	50.7	72.9
5.0– 9.9	21.5	11.2
10.0– 14.9	3.1	6.1
15.0– 19.9	3.1	3.1
20.0– 24.9	4.6	2.7
25.0– 29.9	4.6	0.9
30.0– 34.9	3.1	0.9
35.0– 39.9	1.5	0.4
40.0– 44.9	0.0	0.4
45.0– 49.9	0.0	0.4
50.0– 54.9	1.5	0.2
55.0– 59.9	1.5	0.1
60.0– 64.9	1.5	0.1
65.0– 69.9	0.0	0.1
70.0– 74.9	0.0	0.1
75.0– 79.9	0.0	0.1
80.0– 84.9	0.0	0.0
85.0– 89.9	0.0	0.2
90.0– 94.9	0.0	0.0
95.0– 99.9	0.0	0.0
100.0–104.9	1.5	0.2
105.0–109.9	1.5	0.0

Tab. 4: *A comparison of the % frequency of storms at the experimental site 1984 to 1986 and for Murcia 1951 to 1985.*

gullied bare marls that are so often used to exemplify the desertification problems of this region, though in practice these slopes may not be untypical of large areas of the south-east of Spain.

Finally it is worth considering the distribution of work done by the storms. In the short run, it is tempting to conclude that the two very large storms exemplify the prevailing catastrophic view, that in semi-arid areas all the work is done in a short time by a few very large events. If we define magnitude by the amount of sediment removed by a rainfall event of a given frequency according to our regression results, then we can obtain the form of the work done for events of different frequencies by taking the product of magnitudc and frequency. From fig.3 we can see that this is bimodal. There is a clear peak for small storms of 5–10 mm which, while they do very little work, occur very frequently; and a further peak at 25–30 mm which in the longer term is the most important for sediment movement. Above this the very sharp decline in frequency mitigates against the relatively rare events. The actual data of the period of observation, shown by the pecked line, reveal just how misleading the two large storms are in terms of the overall work done.

6 Conclusions

The Gerlach troughs exhibit high spatial variability in soil erosion and very varying responses to rainfall. The larger catchment shows that rainfall by storms provides a highly significant correlation with sediment yield. If these data are considered at monthly levels by aggregating the sediment yield and precipitation for each month (that is January, February etc.) across the three years the corre-

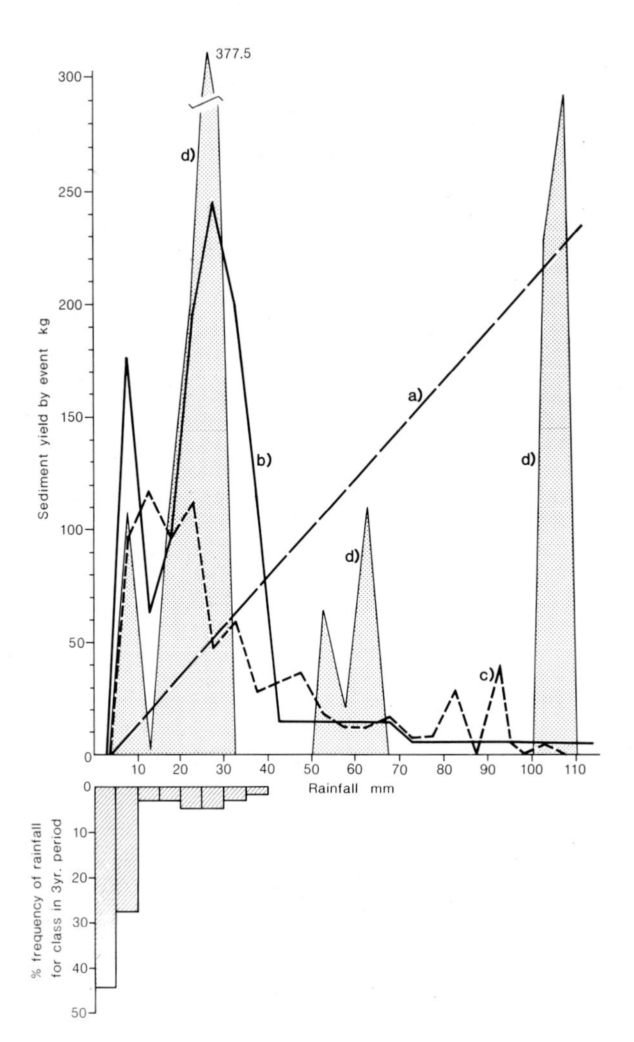

Fig. 3: *Magnitude, frequency and work done according to various assumptions.*

(a) Magnitude for rainfall events according to regression for the collectors on the site.
(b) Product of magnitude (a) and frequency of events as measured over 3 years at the site.
(c) Product of magnitude (a) and frequency of events as measured over 35 years at the Murcia rainfall station. Bar diagram shows frequency for the rainfall events on the site.
(d) Actual sediment yield on the site.

lation between the two is then very high (r=0.93 which is significant at the 99.9% level). The results suggest relatively low annual erosion rates even when the period had fewer small storms and more larger storms than the 35 years period from 1951 to 1986. This is largely attributable to the vegetation cover, the high surface roughness and the slope geometry. Even the two 100 year storms did not raise the erosion rates ot magnitudes comparable to those on badly gullied marls in the adjacent areas. The magnitude and frequency analysis suggests that, at least for these matorral covered hillslopes, the medium sized events are most important in determining the sediment production.

Acknowledgement

We wish to extend our thanks to the British Council and to the Spanish Ministry of Education for their financial support of the field work during this project through the Acciones Integradas scheme.

References

ELIAS CASTILLO, F. & RUIZ BELTRAN, L. (1979): Precipitaciones maximas en España. Ministerio de Agricultura, Monografias **21**, 545 pp.

FISHER, G.C., ROMERO DIAZ, M.A., LÓPEZ-BERMÚDEZ, F., THORNES, J.B. & FRANCIS, C.F. (1987): La produccion de biomasa y sus efectos en los procesos erosivos en un ecosistema mediterraneo semiárido del SE de España. Anales de Biología, **12**. Universidad de Murcia.

FRANCIS, C.F., THORNES, J.B., ROMERO DIAZ, M.A., LÓPEZ-BERMÚDEZ, F. & FISHER, G.C. (1986): Topographic controls of soil moisture, vegetation cover and degradation in a moisture stressed Mediterranean environment. CATENA, **13**, 211–225.

FRANCIS, C.F. (1986): Soil erosion on fallow fields: an example from Murcia. Papeles de Geografía Física, **11**, Universidad de Murcia.

GRUPO DE TRABAJO REGIONAL DEL SEGURA (1985): Estudio sobre la evaluación de recursos hidraúlicos superficiales e hidroeléctricos. **VI**, Erosion Hidrica, Plan Hidrológico Nacional Confederación Hidrográfica del Segura, MOPU.

ICONA (1982): Paisajes erosivos en el sureste español. Ministerio de Agricultura, Pesca y Alimentacion, Proyecto LUCDEME, Monografias **26**, 67 pp.

IMESON, A.C., KWAAD, F.J.P.M. & VERSTRATEN, J.M. (1982): The relationship of soil physical and chemical properties to the development of badlands in Morocco. In: BRYAN, R. & YAIR, A. (Eds.), Badland Geomorphology and Piping. Geo Books, Norwich, 47–70.

IMESON, A.C. & VERSTRATEN, J.M. (1985): The erodibility of highly calcareous soil material from southern Spain. CATENA, **12**, 291–306.

JANSEN, J.M.L. & PAINTER, R.B. (1974): Predicting sediment yield from climate and topography. Journal of Hydrology, **21**, 371–380.

LÓPEZ-BERMÚDEZ, F. (1971): Las precipitaciones en Murcia de 1862 a 1971. Papeles de Departamento de Geografía, **3**, Universidad de Murcia, 171–187.

LÓPEZ-BERMÚDEZ, F. (1986): Evaluación de la erosión hídrica en las áreas receptoras de los embalses de la Cuenca del Segura. Aplicación de la USLE. In: LÓPEZ-BERMÚDEZ, F. & THORNES, J.B. (Eds.), Estudios sobre Geomorfología del Sur de España. Universidad de Murcia, 93–99.

LÓPEZ-BERMÚDEZ, F. (1987): Incidence of soil erosion by water in the desertification of a semiarid Mediterranean fluvial basin: the Segura Basin, Spain. Escuela Tecnica Superior de Ingenieros de Montes, Universidad Politécnica de Madrid (in press).

LÓPEZ-BERMÚDEZ, F., ROMERO DIAZ, M.A., RUIZ GARCIA, A., FISHER, G.C., FRANCIS, C.F. & THORNES, J.B. (1984): Erosión y ecologia en la España semi-arida, Cuenca de Mula, Murcia. Cuadernos de Investigación Geografica, **X**, 113–126.

LÓPEZ-BERMÚDEZ, F. & GUTIERREZ ESCUDERO, J.D. (1982): Estimacion de la erosión y aterramiento de embalses en la cuenca hidrográfica del río Segura. Cuadernos de Investigación Geográfica, **VIII**, 3–18.

ROELS, J.M. (1985): Estimation of soil loss at a regional scale based on plot measurements — some critical considerations. Earth Surface Processes and Landforms, **10**, 587–595.

SCOGING, H.M. (1982): Spatial variations in infiltration, runoff and erosion on hillslopes in semi-arid Spain. In: BRYAN, R. & YAIR, A. (Eds.), Badland Geomorphology and Piping. Geo Books, Norwich, 89–112.

WISE, S.M., THORNES, J.B. & GILMAN, A. (1982): How old are the badlands? A case study from south-east Spain. In: BRYAN, R. & YAIR, A. (Eds.), Badland Geomorphology and Piping. Geo Books, Norwich, 259–277.

YAIR, A., GOLDBERG, P. & BRIMMER, B. (1982): Long term denudation rats in the Zin-Havarim badlands, northern Negev, Israel. In: BRYAN, R. & YAIR, A. (Eds.), Badland Geomorphology and Piping. Geo Books, Norwich, 279–291.

Addresses of authors:
M.A. Romero-Díaz, F. López-Bermúdez
Departamento de Geografia Fisica
Universidad de Murcia
30001 Murcia, Spain
J.B. Thornes, C.F. Francis
Department of Geography
University of Bristol
Bristol, BS8 1SS, England
G.C. Fisher †
Department of Geography
Royal Holloway and Bedford New College
Egham, TW 20 OEX, England

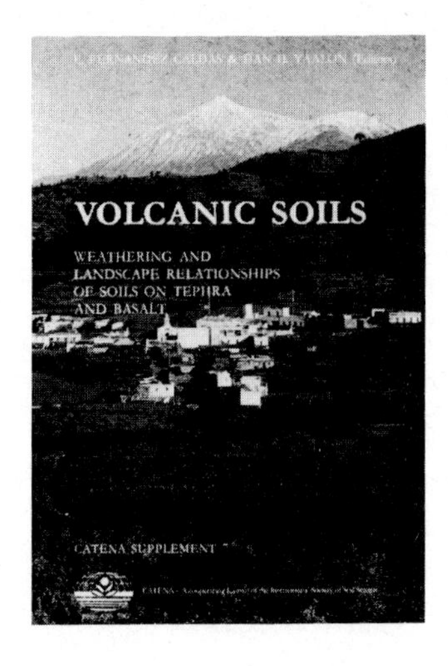

E. Fernandez Caldas & Dan H. Yaalon (Editors):

VOLCANIC SOILS

Weathering and Landscape
Relationships of Soils on Tephra and Basalt

CATENA SUPPLEMENT 7, 1985

Price DM 128,—

ISSN 0722–0723 / ISBN 3–923381–06–9

PREFACE

This CATENA SUPPLEMENT contains selected papers presented at the International Meeting on Volcanic Soils held in Tenerife, July 1984. The meeting brought together over 80 scientists from 21 countries, with interest in the origin, nature and properties of soils on tephra and basaltic parent materials and their management. Some 51 invited and contributed papers and 8 posters were presented on a wide range of subjects related to volcanic soils, many of them dealing with weathering and landscape relationships. Classification was also discussed extensively during a six day excursion of the islands of La Palma, Gomera and Lanzarote, which enabled the participants to see the most representative volcanic soils of the Canary Archipelago under a considerable range of climatic regimes and parent material ages.

Because volcanic soils are not a common occurrence in regions where pedology developed and progressed during its early stages, recognition of their specific properties made an impact only in the late forties. The name **Ando** soils, now recognized as a special Great Group in all comprehensive soil classification systems, was coined in 1947 during reconnaissance soil surveys in Japan made by American soil scientists. Subsequently a Meeting on the Classification and Correlation of Soils from Volcanic Ash, sponsored by FAO and UNESCO, was held in Tokyo, Japan, in 1964, in preparation for the Soil Map of the World. This was followed by meetings of a Panel on Volcanic Ash Soils in Latin America, Turrialba, Costa Rica, in 1969 and a second meeting in Pasto, Colombia, in 1972. At the International Conference on Soils with Variable Charge, Palmerston, New Zealand, 1981, the subject of Andosols was discussed intensively. Most recently the definitions of Andepts, as presented in the 1975 U.S. Soil Taxonomy, prompted the establishment of an International Committee on the Classification of Andisols (ICOMAND), chaired by M. Leamy from C.S.I.R., New Zealand, which held a number of international classification workshops, the latest in Chile and Ecuador, in January 1984. The continuous efforts to improve and revise the new classification of these soils is also reflected in some of the papers in this volume.

While Andosols or Andisols formed on tephra (volcanic ash), essentially characte by low bulk density (less than 0.9 g/cm^3) and a surface complex dominated by activ cover worldwide an area of about 100 million hectares (0.8% of the total land area), th basaltic plateaus and their associated soils cover worldwide an even greater area, frequ with complex age and landscape relationships. While these soils do not generally belo the ando group, their pedogenetic pathways are also strongly influenced by the natur physical properties of the basalt rock. The papers in this volume cannot cover the wide v of properties of the soils in all these areas, some of which have been reviewed at pre meetings. In this volume there is a certain emphasis on some of the less frequently st environments and on methods of study and characterization as a means to advanc recognition and classification of these soils.

The Tenerife meeting was sponsored by a number of national and international o zations, including the Autonomous Government of the Canary Islands, the Institute of American Cooperation in Madrid, the Directorate on Scientific Policy of the Ministry of cation and Science, Madrid, the International Soil Science Society, ORSTOM of Franc ICOMAND. Members and staff of the Department of Soil Science of the University Laguna had the actual task of organizing the meeting and the field trips. In editing the we benefitted from the manuscript reviews by many of our colleagues all over the worlc the capable handling and sponsorship of the CATENA VERLAG. To all those who extended their help we wish to express warm thanks.

La Laguna and Jerusalem,
Summer 1984

E. Fernandez Caldas
D.H. Yaalon
Editors

CONTENTS

CATENA SUPPLEMENT 13 p. 13–35 Braunschweig 1988

FLUVIAL GEOMORPHIC PROCESSES IN SEMI-ARID EPHEMERAL CATCHMENTS IN KENYA AND CANADA

R.B. **Bryan**, Toronto, I.A. **Campbell**, Edmonton, R.A. **Sutherland**, Toronto

Summary

Two morphologically similar semi-arid catchments in Kenya and Canada show the effects of seasonal climatic contrasts and runoff processes in regions of markedly different climate. The Katiorin Catchment, in Kenya, has little seasonal temperature change, but extreme intra- and inter-annual variations in rainfall. The Dinosaur Catchment in Canada experiences profound changes between winter and summer. Most precipitation falls in summer while harsh winters may produce prolonged snow cover. The varying patterns of summer rainstorms, winter snow accumulation and snowmelt produce a wide range of runoff response.

In 1985 four storms with 18–33.6 mm rainfall produced limited runoff in the Katiorin Catchment, with runoff coefficients ranging from 4–16%. By contrast, summer storms in Dinosaur Catchment in 1983, with rainfall ranging from 4.6–40 mm produced runoff coefficients of 28–54%. These differences in runoff response primarily reflect the influence of regolith properties rather than the characteristics and seasonality of rainfall.

Both catchments include a range of regolith materials, but significant portions of each (>40%) are formed of very similar sodic smectitic claystones and mudstones. In Katiorin these units show the most rapid runoff response, but in Dinosaur they are amongst the less responsive units. Stream loads in both catchments are dominated by suspended sediment, with concentrations in the same range, but difference in runoff production results in storm denudation rates which reach 3 t ha^{-1} in Dinosaur, but only 1.5 t ha^{-1} in Katiorin. Despite the high solute production of claystones in Katiorin solutes form a lower propertion of total stream load than in Dinosaur. Bedload transport, however, forms a much more significant component of stream load reflecting marked differences in the material reaching channels from hillslope. In both cases bedload is dominated by sand, but the D_{50} for Katiorin is 1.8 mm, compared with 0.11 mm for Dinosaur. Bedload in Katiorin also includes large armoured mudballs spalled off steep claystone slopes, partially due to tunnel-induced water concentration.

Geomorphologically significant storms are erratically distributed throughout the year in Katiorin, with little clear sea-

ISSN 0722-0723
ISBN 3-923381-13-1

3-923381-13-1/88/5011851/US$ 2.00 + 0.25

sonal concentration, producing an estimated annual denudation of 1.7 mm. In Dinosaur rainstorms between May and September produce an average of 2.1 mm denudation. Winter runoff is extremely variable depending in snow cover and melt conditions. Discharges can exceed peak summer values, with similar suspended loads, but bedload transport is restricted by channel ice. Available data indicate that winter processes may cause 15–30% of the average annual denudation.

Resumen

A pesar de estar situadas en regiones de clima marcadamente distinto, dos cuencas semi-áridas morfológicamente similares de Kenia y Canadá muestran claramente los efectos de los contrastes estacionales en los procesos de escorrentía. En Kenia la cuenca de Katiorin está sujeta a una variación extraordináriamente marcada en la precipitación intra e interanual, si bien los cambios de temperatura son escasos. En la cuenca Dinosaur de Alberta, Canadá, los cambios estacionales de verano a invierno son muy marcados ya que la mayor parte de la precipitación cae en verano, mientras que en invierno, a causa del intenso frío, puede producirse un muy prolongado recubrimiento de nieve. El contraste entre tormentas de verano y fusión de nieve en invierno crea una amplia gama de respuestas hidrológicas.

Durante el año 1985 se estudiaron cuatro crecidas en la cuenca de Katiorin producidas por chubascos que oscilaron entre los 18.2 mm y los 33.6 mm. En ellas la generación de escorrentía fué escasa, con coeficientes que oscillaron entre el 6.7% y el 21.3%. En 1983, en cambio, chubascos entre 4.61 mm y 40.00 mm produjeron, en la cuenca de Dinosaur, coeficientes de escorrentía de 28% y 70% respectivamente. Estas diferencias en la producción de escorrentía son esencialmente el reflejo del carácter del sustrato litológico más que de las características climáticas y la estacionalidad de la precipitación.

Ambas cuencas comprenden varios tipos de regolito pero una gran parte de ellos (>40%) está compuesto por arcillas smectíticas sódicas. En Katiorin estas unidades son las que muestran una respuesta más rápida a la escorrentía, mientras que en Dinosaur por el contrario son las que tienen una respuesta más baja. La carga de sedimentos es en ambas cuencas predominantemente en suspensión, con concentraciones similares, pero las diferencias en la escorrentía determinan tasas de denudación que llegan a 3 t ha^{-1} en Dinosaur y sólo a 1.5 t ha^{-1} en Katiorin. A pesar de que las arcillas producen en Katiorin una cantidad elevada de solutos, éstos constituyen una proporción más pequeña del total de la carga que en Dinosaur. El transporte de fondo, no obstante, es un componente mucho más significativo de la carga fluvial puesto que refleja notables diferencias en el material que llega a los canales desde las vertientes. En ambos casos la carga es predominantmente arena, pero el D_{50} es en Katiorin de 1.8 mm mientras que en Dinosaur es de 0.11 mm. En Katiorin se encuentran además grandes bolos de barro seco procedentes de hundimientos causados por procesos de sufusión en las vertientes.

En Katiorin los chubascos geomorfológicamente significativos están distribuídos de manera errática a lo largo del año, con concentraciones estacionales poco claras, y se ha estimado que producen una denudación anual de 1.7 mm. En la cuenca de Dinosaur los chubascos que tienen lugar entre mayo y septiembre producen una denudación media de 2.1 mm. La escorrentía es en invierno extremadamente variable puesto que depende de la cobertura nival y de las condiciones de fusión. Los caudales pueden superar los máximos estivales pero aunque las cargas en suspensión son similares el transporte de fondo queda restringido por el hielo que ocupa el canal. Los datos de que se dispone indican que los procesos que tienen lugar en invierno pueden ser la causa de un 15–30% de la media anual de la denudación.

1 Introduction

In humid regions the influence of strong seasonal climatic contrasts on fluvial geomorphic processes is shown by data from many instrumented catchments. By contrast, in arid and semi-arid regions, despite a number of investigation (e.g. SCHICK 1977, YAIR et al. 1980, BRYAN & CAMPBELL 1982, 1986) few data are available. Many arid and semi-arid regions experience marked variations in precipitation and stream discharge, but these often appear to be random rather than seasonally-distributed events, and so the geomorphic influence of seasonal contrasts has received little attention.

In this paper fluvial geomorphic processes in two instrumented semi-arid ephemeral catchments with very different seasonal climatic regimes are examined. The Katiorin Experimental Catchment, established in 1984 in the Baringo District, Kenya, provides information on the geomorphology of a small ephemeral drainage system in the tropics. The Dinosaur Experimental Catchment was established in 1981 in Dinosaur Provincial Park, Alberta, Canada. Research in the latter has focussed primarily on the geomorphic influence of summer rainstorms, but the region has harsh winters and about one-third of the average annual precipitation is snow. The two catchments were established for independent research projects, but similarities in size, morphometry, research methodology and some aspects of lithology provide an unusual opportunity for direct comparison of fluvial geomorphic processes in tropical and temperate regions. There is also an opportunity to compare processes in a rapidly-eroding natural temperate area with a tropical region where erosion rates appear comparable, but where erosion has been human-induced.

2 Study Areas and Catchment Characteristics

2.1 Katiorin Experimental Catchment

The Katiorin Catchment is situated to the west of Lake Baringo at 947 m asl, on the eastern edge of the Tugen Plateau, one of the minor fault blocks which fringe the main western wall of the Central Rift Valley, Kenya, which rises to over 2500 m asl in the adjacent Kamasia Range (fig.1). The catchment drains an area of 31.1 ha through a third order channel (tab.1) into the Kapthurin River, one of large ephemral channels which rise on the Kamasia Range and drain onto the extensive alluvial and lacustrine Njempt Flats which separate the Tugen Plateau from Lake Baringo.

Regional geology is complex, being dominated by Miocene volcanics intercalated with pene-contemporaneous al-

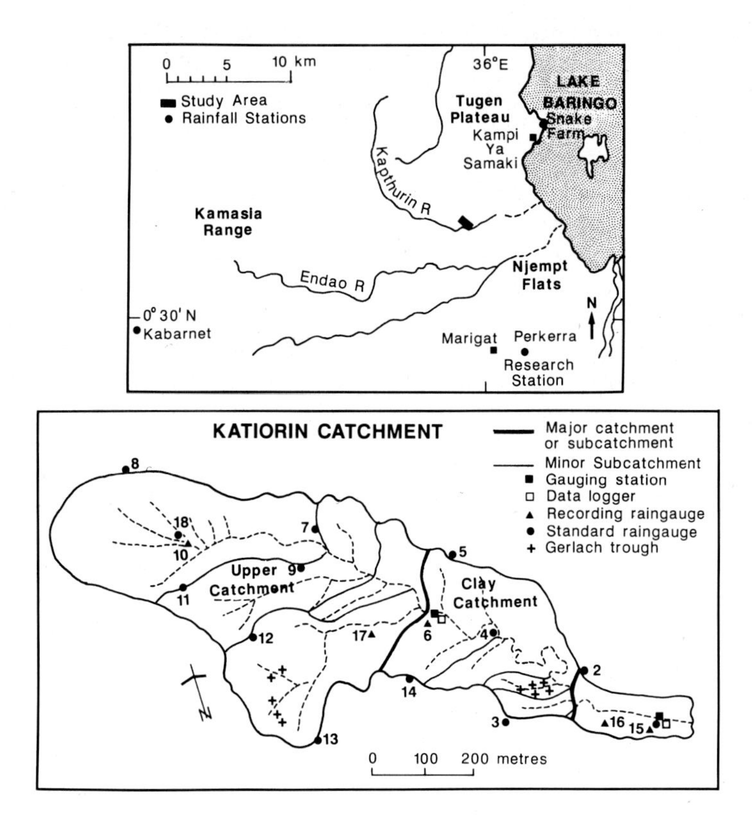

Fig. 1: *Vicinity of Lake Baringo, Kenya and map of Katiorin Experimental Catchment.*

Characteristic	Katiorin	Dinosaur	Difference
Order	3	4	
Area (ha) (A)	31.1	33.7	7.6
Length (km) (Bl)	1.27	1.14	11.4
Elongation (Se)	0.49	0.57	14.0
Perimeter (km) (P)	3.27	3.11	5.0
Compactness (Sc)	2.74	2.29	16.4

Bl = straight line distance from lower gauging station to furthest point on drainage divide.
$A_2^{0.5} \times 2/Bl\times^{0.5}$
$P - A/4\times$ (GARDINER 1975)

Tab. 1: *Comparison of selected morphometric characteristics of Katiorin and Dinosaur Catchments.*

Photo 1: *Upper portion of Katiorin Catchment, showing non cohesive rilled slopes and calcareous tuff caprock.*

Unit	Sand (%)	Silt (%)	Clay (%)	Electrical Conductivity (1:99 Soil:Water; μS cm^{-1}) at 25°C			Exchangeable Bases (meq/100g)			
				30 min	60 min	240 min	Na^+	Ca^{++}	Mg^{++}	K^+
Katiorin Catchment										
Red Clay	8	22	70	247	302	380	39.08	21.54	2.16	1.89
Middle Silts and Gravels										
(mid-catchment)	83	14	3	90	188	260				
(upper catchment)	92	7	≤1	55	68	90				
(upper catchment)	85	14	≤1	38	47	113				
(upper catchment)	91	8	≤1	68	75	95				
(unit terminology after TALLON 1978)										
Dinosaur Catchment										
Yellow Mudstone	5	49	46	94	145	214	5.47	0.38	0.11	0.06
Grey Mudstone	7	37	56	98	112	148	11.36	2.14	0.59	0.21
Sandstone	45	33	22	172	192	n.a.	1.85	0.12	0.02	0.02
Pediment	36	55	9	n.a.	n.a.	n.a.	2.02	0.42	0.07	0.02

Tab. 2: *Physical and chemical properties of selected lithological units.*

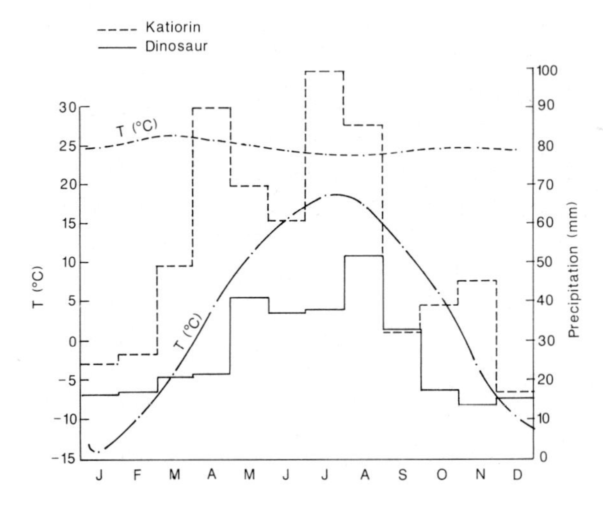

Fig. 2: *Average climatic data for Perkerra Experimental Station, Baringo, and Brooks Horticultural Station, Alberta.*

luvial and lacustrine deposits, much disturbed by tectonism (KING 1970 CHAPMAN et al. 1978). The catchment itself is incised into fluviolacustrine silts, clays and gravels intercalated with calcareous tuffs and pumices of the Pleistocene Kapthurin Formation (TALLON 1978). The catchment consists of an upper sector (fig.1) dominated by poorly consolidated silts and sands forming densely-rilled steep slopes below the calcareous tuff plateau (photo 1), and a central and lower portion in which a marked constricted meander zone is incised into red and black sodic smectitic claystones (tab.2). These are extensively disrupted by desiccation cracks and tunnel erosion. In its lowest section the channel is incised to a depth of 1 m into an alluvial terrace of the Kapthurin.

The Lake Baringo basin is uniformly hot throughout the year, and semi-arid, with a mean annual rainfall at Perkerra (fig.1) of 634 mm and potential evapotranspiration of 2576 mm giving a moisture deficit ratio of 4.06. There is a steep transition with elevation to the summit of the Kamasia Range where the mean annual rainfall exceeds 1445 mm. Mean monthly rainfall figures show quite pronounced concentration in April and July/August (fig.2) and there is a high seasonality index (wettest 3 months/ driest 3 months: 4.19 mm). Detailed analyses of rainfall characteristics (RODGERS et al. 1982, ROWNTREE 1986) show that this pattern is often obscured by erratically distributed heavy storms which can produce more than 10 per cent of the mean annual rainfall in a single storm. Most rain falls in convective storms in the early evening. More than 80% of the rain falls in storms <10 mm, but these account for only 36% of the annual rainfall. Peak recorded 30 minute rainfall intensity was 80 mm h^{-1} and ROWNTREE (1986) predicts long-term values of 100 mm h^{-1}.

Vegetation shows marked elevation zonation, but the Tugen plateau is dominated by *Acacia* thorn shrub, with poor ground vegetation. Vegetation degradation and soil erosion followed land-use changes and overstocking in the 1920's, exacerbated by recurrent droughts (ANDERSON 1981). Soil degradation is extreme and at most locations only truncated sub-soils remain. Recent estimates of contemporary denudation rates of 2.2 and 2.3 mm yr^{-1} have been made for the neighbouring Chemeron and Endao catchments (BARINGO PILOT SEMI-ARID PROJECT 1983). These rates seem consistent with other erosional evidence, and sedimentation in Lake Baringo.

Catchment instrumentation (fig.1) includes automatic and recording raingauges, checked after storms, two gauging stations with pressure transducer stage recorders linked to data loggers, one automatic water/suspended sediment sampler (at the lower station) and Gerlach troughs. Automatic sampling is supplemented by manual water and suspended sampling whenever possible, and by bedload sampling with a hand-held Helley-Smith sampler. Erosion pin grids have been set up throughout the catchment, and runoff response of lithological surfaces has been measured with an Amsterdam-pattern sprinkling infiltrometer.

2.2 Dinosaur Experimental Catchment

The Dinosaur Experimental Catchment is situated in an extensive badland zone along the Red Deer River in semi-arid southeastern Alberta (fig.3). These badlands formed in lagoonal, shallow water or deltaic sediments of the Upper Cretaceous Judith River Formation (KOSTER 1984). Lithology is diverse (tab.2) but is dominated by sodic, smectitic mudstones and claystones and densely-rilled siltstones and fine sandstones (photo 2). Virtually all units are highly erodible and denudation rates are amongst the highest reported from natural undisturbed areas. CAMPBELL (1974, 1981) reported average denudation rates of 3 mm yr^{-1} and peak values reaching 13 mm yr^{-1}. The badlands formed after Wisconsinian deglacation about 14000–15000 B.P. Denudation rates varied through the Holocene, but appear to have been highest during the immediate post-glacial period. Denudation almost ceased after extensive loess deposition around 5200 B.P. (BRYAN et al. 1984) but rates now appear to be gradually increasing as the loess cover is progressively stripped.

Southeastern Alberta has intensely cold, long winters, but short hot summers give an average potential evapotranspiration of 570 mm. As much of the limited annual rainfall falls during the summer (fig.2), moisture deficits are severe, as reflected by the natural shortgrass prairie vegetation. Moisture deficit and seasonality indices are both lower than at Lake Baringo (1.66 and 2.83 respectively). Summer rain typically falls as short, relatively intense (30 mm h^{-1}) convective storms or as more prolonged low intensity (1-5 mm h^{-1}) storms. High intensity rainfall is much less frequent than at Baringo, and the peak intensity of 80 mm h^{-1} for 30 minutes noted by ROWNTREE (1986) at Baringo has a 100 year return frequency at Dinosaur (tab.3). Approximately 30% of the annual precipitation falls as snow, but distribution is extremely varied. Snow may accumulate for release in one major spring melt, but is usually depleted by sublimation or chinook melts and sometimes, no snow-pack persists until spring. Spring snowmelt therefore varies greatly in magnitude and may range from late March to early May, significantly affecting melt-induced discharge and stored regolith moisture at the start of the rainfall season.

The Dinosaur Catchment is a fourth-

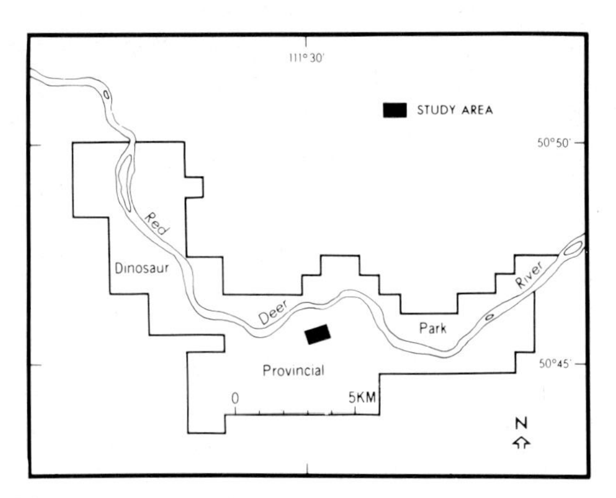

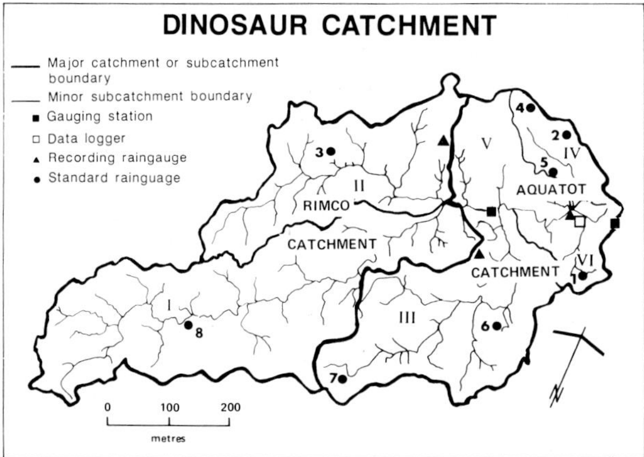

Fig. 3: *Vicinity of Dinosaur Provincial Park, Alberta, and map of Dinosaur Experimental Catchment.*

	Baringo (Ewaso Ngiro-Narok)				Dinosaur (Brooks)			
Years	5	10	25	100	5	10	25	100
15 min	90	110	130	160	53	66	83	109
30 min	60	70	80	100	39	50	64	84
1 h	38	44	50	63	27	35	45	59
2 h	25	27	30	38	16	20	25	33
6 h	9	11	13	16	6	7	9	11

Sources: Ewaso Ngiro-Narok (MINISTRY OF WATER DEVELOPMENT 1978), Brooks (ATMOSPHERIC ENVIRONMENT SERVICE 1978)

Tab. 3: *Rainstorm return frequency data (mm h^{-1}).*

Photo 2: *Tunnel-induced collapse of smectitic red clays into lower channel, Katiorin Catchment.*

order system which resembles the Katiorin Catchment closely in morphological characteristics (tab.1). It includes examples of all major badland lithologies and surface in six significant sub-catchments drained by a well-defined channel network. Instrumentation (fig.3) which includes manual and recording raingauges, two automated gauging stations and two automated water/suspended sediment smaplers, has been described in detail by HONSAKER et al. (1984) and BRYAN & CAMPBELL (1986). Automated stream sampling is supported by manual sampling and bedload sampling with Helley-Smith samplers. Catchment runoff processes have been monitored since 1981, together with shorter studies of the ubiquitous tunnel erosion which can transport as much as 25% of the catchment discharge (BRYAN & HARVEY 1985). Numerous short-term experiments have also been carried out on surface response to rainfall, solute and sediment entrainment (BRYAN & HODGES 1984, BRYAN et al. 1984, BOWYER-BOWER & BRYAN 1986).

3 Catchment Response

3.1 Katiorin Catchment

Observations started in 1984, the driest year on record since records started at Perkerra in 1958, but neither of the two minor storms during the observation period produced catchment channel flow. In 1985, 6 storms occurred during observations and 4 produced channel flow. Since continuous monitoring started in February, 1986, 12 flow-producing storms have been observed. The precise threshold for flow obviously varies with storm intensity and antecedent moisture condi-

Date	Rainfall (mm)	Duration (h)	Average Intensity (mm h^{-1})	Recurrence Frequency (no/yr)	Runoff Coefficient (%)	Load Distribution Sus. (%)	Bed	Sol.	Total Load (kg)	Denudatio (t/ha (mm)
Katiorin Experimental Catchment										
18.6.1985[1]	18.2	2.0	9.1	16	4	71.6	27.0	1.0	13040	0.38 (0.04)
12–13.7.1985[2]	32.2	6.0	5.4	4	10	n/a	n/a	n/a	n/a	n/a
24–25.7.1985[3]	24.9	5.5	4.5	7	16.5	80.0	19.0	1.0	45982	1.48 (0.15)
Dinosaur Experimental Catchment										
18.5.1983[4]	4.6	1.8	2.6	12	28.7	98.7	0.96	0.34	24325	0.72 (0.07)
2.–3.7.1983[5]	40.0	24.0	1.7	4	54.0	97.3	1.84	0.82	103095	3.05 (0.31)

[1] Bedload estimated from filling rate of bedload trap.
Solute load calculated using regression of solute load against electrical conductivity developed for Dinosaur Catchment (SUTHERLAND 1983)
[2] Observer unable to reach catchment to sample sediment or solute load.
[3] Load distribution based on proportional distribution from June 18 storm.
[4] Load distribution based on pumping sampler suspended load measurements; bedload collected with Helley-Smith sampler; solute from regression equation (SUTHERLAND 1983)
[5] Storm occurred in three phases; data for complete storm are shown. Bedload estimated on proportional basis from Rimco gauging station due to loss of sampler at main station - probably conservative estimate.

Tab. 4: *Summary of rainstorm, runoff and sediment transport.*

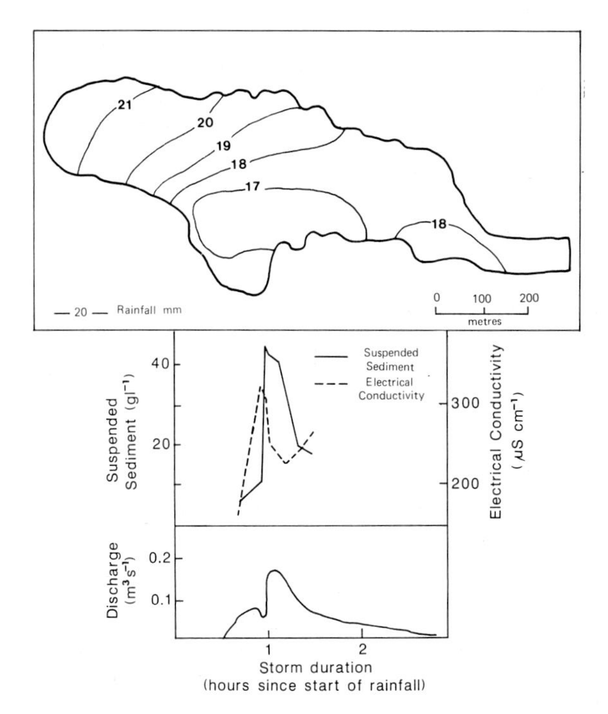

Fig. 4: *Storm of June 18, 1985, Katiorin Catchment.*

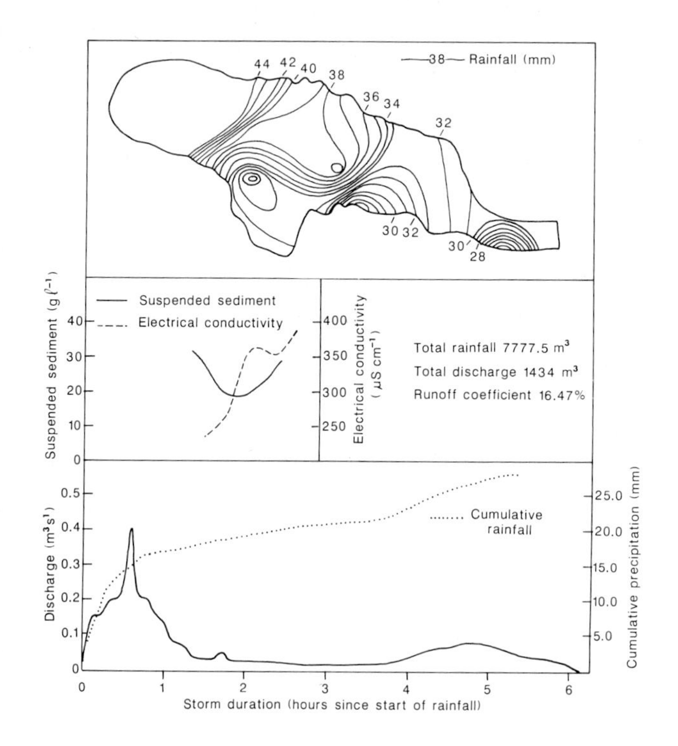

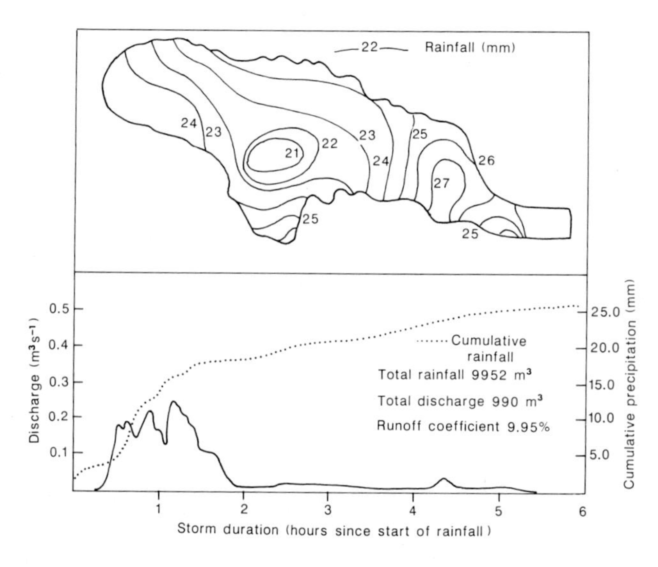

Fig. 5: *a: Storm of July 12–13, 1985, Katiorin Catchment. b: Storm of July 24–25, 1985, Katiorin Catchment.*

tions. ROBERTS (1985) suggests that a threshold rainfall of 15 mm produces significant rainfall on the Njemp Flats while ROWNTREE (1986) used a general threshold of 10 mm which agrees well with STOCKING & ELWELL's (1976) threshold of 12.5 mm for erosive rainfall in East Africa. These thresholds can now be refined on the basis of the stormflow record and data from sprinkling infiltrometer experiments in the catchment.

The overall storm record is not yet sufficient to document all aspects of catchment response as it includes no storms of very high intensity. It does, however, include a good array of storms in the 10–40 mm range which account for 16.2 per cent of all storms. Data sets are not complete for all storms as it was not always possible for observers to reach the catchment during storms. The 1985 storms for which data are shown in tab.4 and figs.4 and 5 appear to be typical of the most common flow-producing events. All were caused by convective activity in the early evening and were extremely localized. Magnitude-duration-return frequency curves are not available but the estimates in tab.4 are based on ROWNTREE's (1986) analysis. Contrary to general expectations of tropical rainfall, average intensities were low, ranging from 3.1–9.1 mm h^{-1}. Highest peak intensities were 60 mm h^{-1} for 4 minutes on July 24–25 and 36 mm h^{-1} for 5 minutes on July 12–13.

The most conspicuous features of catchment runoff response are the very low runoff coefficients, particularly on June 18 (fig.4). Although the preceding two weeks, and the preceding year were dry, more than 400 mm of rain fell in Jan–May, 1985, the fourth highest total for the period since records started. The influence of antecedent moisture can be seen in the first 18 mm of rain in the storm of July 24–25 (fig.5b) which followed storms on July 12–13 (fig.5a) 19 and 23–24, and produced a runoff coefficient of 18.4%.

Sprinkling infiltrometer tests carried out on various surfaces at intensities from 25.4–36 mm h^{-1} cast some light on flow patterns in the catchment. On steep non-coherent silty and sandy slopes in the upper catchment wetting depths ranged from 3–5.2 cm and 35 minutes (21 mm) of rain was required to produce ponding. On red and black clays in the lower catchment general wetting depths (not in desiccation cracks) were 2–2.5 cm and runoff started after 11–20 minutes (6–12 mm) of rain. These tests confirm observations that initial storm flow at the main gauging site is derived only from the clay catchment and lower channel areas. During this initial period suspended sediment loads are low (fig.4) and are dominated by clays. As channel flow velocities in this period reached 0.77 m s^{-1} this clay dominance must reflect scarcity of silt and fine sand on surfaces in the lower catchment. Very low surface erosion rates recorded at Gerlach troughs on clay surfaces indicate that water and suspended clay in early storm flow must be contributed by tunnels.

Arrival of flow from the upper catchment is shown in the June 18 storm (fig.4) by the marked increase in discharge after storm duration of 65 minutes. Suspended sediment concentrations increased from 10.7–45.3 g l^{-1} in 3 minutes without significant changes in flow velocity, due to increased availability of silt and fine sand. The latter concentration is in the medium range (BEVERAGE & CULBERTSON 1964) and is quite comparable to concentrations in the Dinosaur

Catchment, but because of the low runoff yield, storm sediment yield on June 18 was only 0.03 kg m^2.

The patterns of electrical conductivity in storm flow closely reflect contributing lithologies. The clays have a high capacity to release solutes, as shown by the equilibrium value of 380 S cm^{-1} for the Katiorin red clay in laboratory tests (tab.2). As a result of clay dissolution the electrical conductivity of flow increased rapidly early in the storm, but once flow was diluted by contributions from the upper catchment where surfaces have much lower solute release potential (as low as 90 S cm^{-1}, tab.2), it dropped sharply. The June 18 storm did not last sufficiently long to show subsequent changes in conductivity, but the limited data from the July 24–25 storm (fig.5b) suggest that the dilution effect probably disappears in more prolonged storms. This is consistent with the prolonged wetting necessary to produce maximum solute release on smectite clays (tab.2).

Bedload transport showed several interesting features. The record of instantaneous transport rates based on Helley-Smith samplers is still limited but indicates high rates reaching 164 kg min^{-1} for the complete channel cross-section. As in Alberta (BRYAN & CAMPBELL 1986) there was some concern that these samples might undercatch, so a 0.9 m^3 bedload trap was dug below the main station in 1986. In each of 5 storms this filled and aggraded, indicating minimum transport of 1300 kg. In at least one storm it filled in the first 20 minutes of flow, so in most storms the total bedload transport must be very much higher. Bedload is dominated by coarse sand grade material with a D_{50} of 1.8 mm, but this consists partly of red clay aggregates. A conspicuous feature of the catchment is the collapse of material from steep clay slopes in the meander zone into the channel during and immediately after storms (photo 3). Some of these failures are induced by tunnel flow during storms but most are dry ravelling as the highly swollen red clays dry and shrink after storms. Clay blocks reaching the channel vary greatly in size, but some exceed 0.5 m in diameter. Some fragments survive several storms, gradually disintegrating in place, but many roll downstream to form armoured mudballs (photo 4). These are abundant in the channel below the meanders, and can persist below the junction with the Kapthurin. Most are much larger than similar features described from other areas (BALL 1940). A random survey of mudballs moved during storms in early 1986 gave a mean diameter of 0.2 m, and the largest moved was 0.39 m diameter with a mass >20 kg. In 3 channel sections a total of 324 mudballs were counted. Mean transport distance could not be determined, but maximum and minimum distances were 375 and 100 m. If the mid-point of 240 m is a reasonable reflection of mean transport distance, this would represent a single event bedload movement of 1200 grain diameters. This greatly exceeds usual estimates (LEOPOLD & EMMETT 1981) and is of the same order as POESEN (1987) noted for rill flow.

While precise quantities are not yet known, mudball transport is evidently a significant mode of sediment transport. Observations suggest that most mudballs move only in the first flow event after they reach the channel, becoming saturated and embedded in sandy bed load during recession. They subsequently break down **in situ** yielding the coarse

Photo 3: *Characteristic assemblage of slope forms and lithological units, Dinosaur Catchment. Lower silty micro pediments, intermediate mudstone, and upper rilled sandstone.*

Photo 4: *Armoured mudballs in lower channel, Katiorin Catchment.*

clay aggregates found in bed load samles. It is notable that even minor storms produce mudball transport and mudballs up to .08 m in diameter moved in the June 18 storm even in flows <0.05 m in depth.

3.2 Dinosaur Catchment

3.2.1 Summer Processes

Between May and September 16–20 channel flows typically occur in Dinosaur Catchment. These include minor events caused by low intensity (1–5 mm h^{-1}) frontal rainstorms and more major events caused by convective storms of moderate (20–30 mm h^{-1}) intensity, usually of brief duration. High intensity rainfall is rare, the peak recorded being 80 mm h^{-1} for 6 min at the start of a prolonged storm in 1981. Although catchment response has been monitored since 1981, very few storms occurred during 1984–85. Storms from 1981 and 1982 have been analysed by BRYAN & CAMPBELL (1982, 1986) so storms from 1983 have been used for this study.

The key factors dominating catchment runoff are storm characteristics and the differential response of surfaces. Basically these fall into high, moderate and low runoff-yielding categories. Siltstones, sandstones, silty micropediments and channels all have low infiltration capacities and generate runoff swiftly, producing high runoff coefficients and sediment entrainment, but notably low solute production (BRYAN et al. 1984). Mudstone surfaces vary in response depending on desiccation crack density, sealing rate and antecedent moisture, and usually flow only during storms of moderate-high intensity or prolonged duration. Water and sediment volumes are quite large and solute release is greater than on high-yielding surfaces and increases with storm duration due to progressive mustone dissolution (tab.2). Finally, grassed, loess-mantled surfaces have very high infiltration capacities (>6 mm h^{-1}, HODGES 1984) and yield runoff only in the most intense or prolonged storms. The influence of surface character on catchment response is clearly most pronounced in storms of low intensity or duration, especially if these occur in dry antecedent conditions. A final factor of importance is tunnel flow, which can follow complex routeways to channels. Discharge varies greatly with subsurface moisture conditions but tunnel flows typically have significantly higher suspended sediment and solute concentrations than channel flows (BRYAN & HARVEY 1985).

The storm of May 18, 1983 (fig.6) typifies catchment response to low intensity frontal rainfall. It followed two weeks without rain, lasted 1.75 h and produced an average of 4.6 mm rain evenly over the catchment. Peak discharge occurred after one hour and was almost simultaneous at both gauging stations. Despite dry antecedent moisture conditions, 459 m^3 discharge were generated giving a catchment runoff coefficient of 28.7%. Runoff did not occur on grassed loessal surfaces which form 26% of the catchment and rainfall simulation experiment data (BRYAN et al. 1978) show that little runoff would have occurred from mudstones. This is confirmed by low electrical conductivities of flow which declined during the storm.

Little sediment transport occurred, with a suspended yield of 0.07 kg m^2 for the complete catchment and 0.04 kg m^2 for the Rimco catchment (fig.2). Total bedload transport at the main station was 223 kg with a D_{50} of 0.11 mm, and

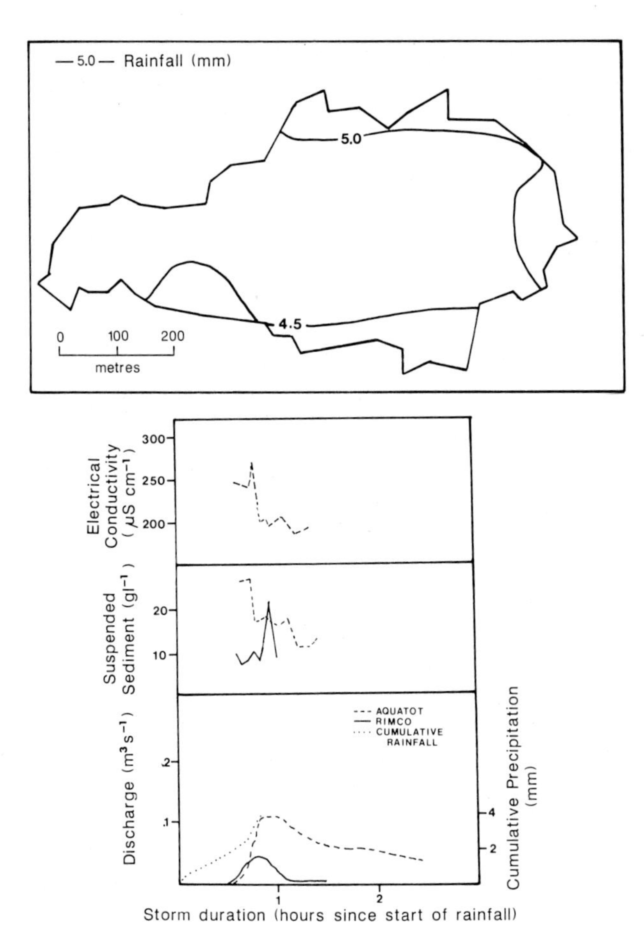

Fig. 6: *Storm of May 18, 1983, Dinosaur Catchment.*

at the Rimco station no bedload at all was collected due to entrappment in the rough channel above the flume. Solute transport formed less than 0.5% of the total load. Some tunnel flow was observed near the main station by the end of the recession but it was insufficient to affect the overall solute concentration.

On July 2 and 3, 1983, a series of localized convective storms produced an average of 40 mm of rain in 24 hours. This was not equally distributed (fig.7) and high-yielding surfaces in sub-catchment 4 (fig.3) received 40% more rain than the loessal grassland in sub-catchment 1. Three millimeters of rain in the initial 5 hours produced no flow but wetted surfaces so the response was rapid when rainfall intensity increased to a peak of 26 mm h^{-1} over 15 minutes (fig.7). Flow occurred from all surfaces except grassland, giving a catchment runoff of 46%. The hydrograph shows the extreme sensitivity of the catchment to minor variations in rainfall intensity once it is saturated. This is also shown by the enhanced response during subsequent phases of the storm. In the second phase, 3 h later, 5.5 mm of rain fell at an average intensity of 4.8 mm h^{-1}, yet the runoff coefficient

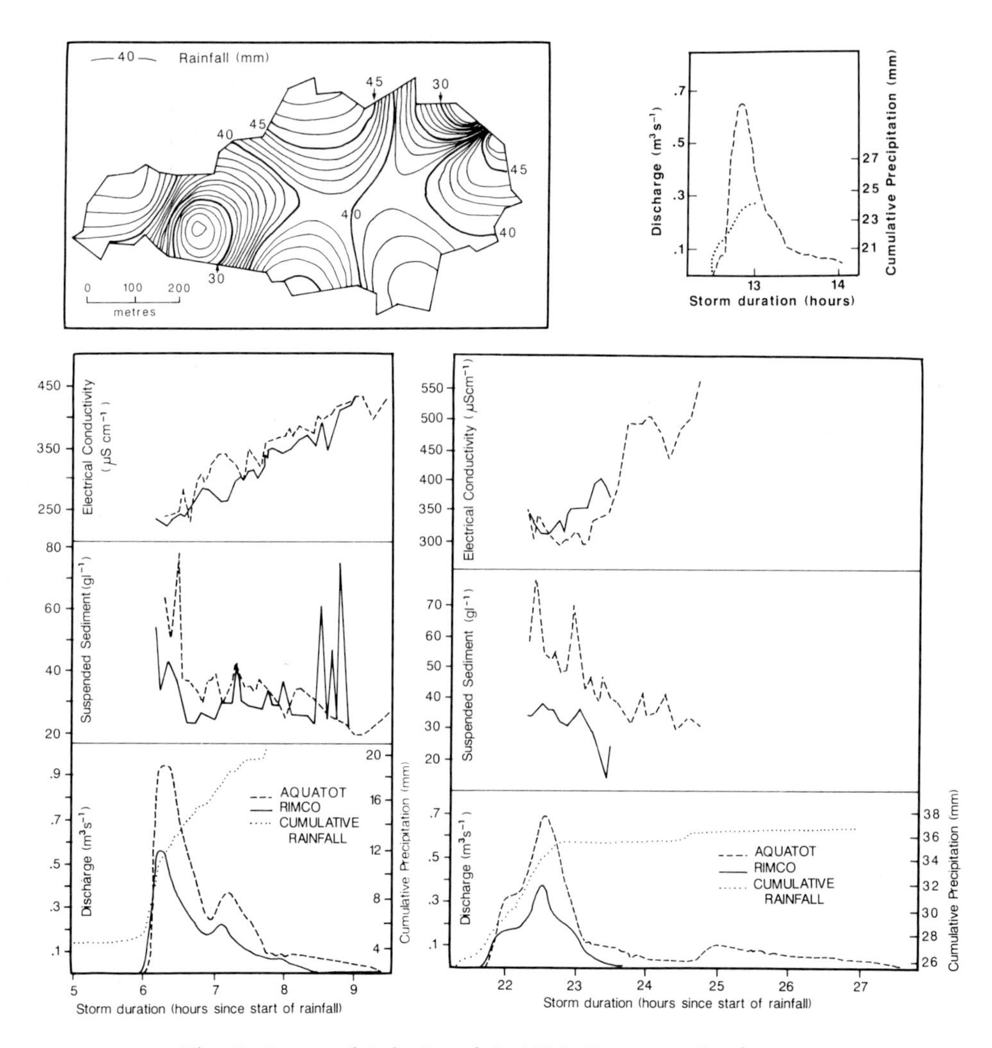

Fig. 7: *Storm of July 2 and 3, 1983, Dinosaur Catchment.*

was 61.5%. In the third phase, 8 h later, 12 mm of rain at an average intensity of 6.7 mm h^{-1} gave a runoff coefficient of 72 %. The average runoff coefficient for the complete storm was 54%.

Sediment transport was substantial during the complete storm sequence, especially during initial flow when rainfall was most intense, and suspended load for the complete catchment was 100350 kg (0.30 kg m^2). Interestingly, the yield for the Rimco catchment was only sightly lower at 0.26 kg m^2 despite the large area of grassland, reflecting the degree to which this is compensated for by high-yielding siltstones and silty micro-pediments. Suspended sediment concentration (fig.7) shows the importance of these areas. The smectite clays released from mudstone surfaces are extremely fine and remain suspended for long time periods regardless of flow velocity. The close linkage of suspended sediment concentration to discharge, particularly at the main station, shows that most suspended sediment is relatively coarse, being derived from siltstones, sandstones and micropediments. High concentrations at the Rimco station during recession probably reflect belated clay-rich contributions from tunnel flow.

Electrical conductivity was not closely related to discharge but increased progressively with storm duration. This contrasts with the May 18 storm in which most flow was derived from sandstones, siltstones and micropediments on which solute release is limited following initial flushing of precipitation. The July 2–3 storm shows the effect of increasing release of solutes with storm duration due to mudstone dissolution and the increasing contribution of solute-rich tunnel flow. While the relative contribution of solutes increased with storm duration, it remained, at 0.82% (tab.4) an insignificant component of the total load.

Only sporadic bedload data are available for the main station as the bedload sampler was lost during peak discharge in the first part of the storm. At the Rimco station bed load yield was 972 kg (1.84% of the total load at that station). Bedload transport rates at the main station are generally much higher than at the Rimco station (BRYAN & CAMPBELL 1986) so it is likely that bedload formed at least as high a proportion of the total load. In general, the low proportion of total load transported as bedload apparently reflects the small size of most material delivered to the channels. As the average D_{50} for all storms is 0.11 mm, at moderate to high discharges most material is carried in suspension. This is consistent with a bed load transport pattern in which concentrations are highest during rising or falling stages, with marked diminution at peak discharge. A similar pattern has been described for an ephemeral stream in northern Kenya by FROSTICK et al. (1983).

3.2.2 Winter Processes

Sub-zero temperatures prevail between late October and early April and an average of 100 mm water equivalent of precipitation falls as snow. Factors influencing the snow climate of Dinosaur Provincial Park have been described by HARTY (1984). Four snow properties may significantly affect catchment processes: duration and depth (water equivalent) of snowpack, and rate and frequency of melting. The first two determine the extent to which underlying surfaces are protected from frost, while the rate and frequency of melting determine the magnitude of snowmelt discharge. Snow cover typically starts in late November and reaches maximum accumulation by mid-February with an average depth of 10–13 cm (HARTY 1984).

High winds throughout the winter cause significant redistribution, drifting and sublimation. Drifting is enhanced by the topography and reflects the dominance of southwesterly winds. Snow may accumulate without much melting until late March or April before one major thaw, but usually several thaws occur in late winter due to chinooks. These may produce a number of flows and reduce discharge associated with the final thaw.

While freeze-thaw activity occurs during the winter it is not of great geomorphic significance. Regolith is usually dry and little affected by frost. Even when materials are moist, most do not fall within the frost-susceptible size range in which ice lensing would occur. The only units in this range are sandstones, siltsones and silty micropediments, but their infiltration characteristics are such that moisture penetrates only very shallow surface layers (HODGES & BRYAN 1982).

Between late November and mid-March snow cover is usually adequate to protect underlying surfaces from frost disturbance. However, southwest slopes may lose snow by wind redistribution and radiation melting, and experience localized disturbance. HARTY (1984) described minor thaw-induced slumping and solifluction moving material to channels and this seems to be of greater geomorphic significance than the effect of frost on regolith erodibility. In any case the erodibility of these materials is so high that frost can only be of slight additional importance. Material transported to channels by slumping or solifluction may be removed by flow during a minor thaw, or may collect for removal during the final thaw or summer rainstorms. In most cases removal apparently occurs before the summer season.

Variations in snow depth and distribution, and in the timing and intensity of thaws could produce a wide range of snowmelt discharges. HARTY (1984) noted that the lithologically-controlled partial area contributions which dominate summer runoff (BRYAN et al. 1978) disappear during the winter and are replaced by partial area contributions determined by snow accumulation patterns. HARTY (1984) measured part of a major melt discharge between February 16 and 20, 1982. Measurements at the main station for the last 53 h of flow showed discharge of 7632 m^3. this is a conservative estimate of melt discharge as it omits flow during the first 48 h, but nevertheless it exceeds the maximum total discharge observed during summer storms. The average discharge was only 0.04 m^3 s^{-1}, and the peak discharge was 0.09m^3 s^{-1}, below the peak of a moderate intensity summer storm. Suspended sediment concentrations were comparable to those in similar summer flows, but the channel bed was protected from scouring by channel ice and there was no bedload transport. By the time flow occurred many surfaces were snow-free so the suspended sediment represents a highly-efficient removal of material from a very restricted contributing area consisting primarily of snow-filled rill channels (HARTY 1984).

Several chinook-induced flows of the type described occur in most winters. The chief geomorphic effect is removal of materials from hillslopes to channels. HARTY (1984) observed little accumulation in channels, but this would vary with the meltwater volumes. In winters with sparse snow cover few flows occur to flush sediment through the channel. This would contribute to channel aggradation, as recently observed in the lower catchment after the low snow winters of 1983–84 and 1984–85. Channel ice normally protects the bed from scouring even during major flows, but significant scouring could occur following rapid melting of a complete winter snowpack in one springtime. In this extreme case channel ice would be minimal and discharge could exceed the largest summer rainstorms.

Little information is available on the effect of winter processes on tunnel development. HARTY (1984) observed significantly deeper penetration of moisture beneath snowpacks than during summer rains. In smectitic mudstones this would enhance swelling, shrinkage and desiccation cracks and could encourage tunnel development. In late winter snow accumulates in tunnels where it is protected from direct radiation and can persist until July. By maintaining wet antecedent conditions in tunnels this could significantly increase tunnel discharge during early summer storms.

4 Discussion and Conclusions

In both catchments, diverse lithology, variations in storm characteristics and irregular topography, together with complex tunnel networks result in highly variable storm response. The Dinosaur Catchment produces significant discharge from most storms but the size and nature of the contributing area varies greatly with storm characteristics. Runoff coefficients vary between 20 and 70% depending on storm conditions. In the Katiorin Catchment, thresholds for runoff initiation are higher, and runoff coefficients correspondingly lower, reflecting a smaller proportion of high-yielding surfaces, despite the greater frequency of high intensity rainstorms (tab.4).

In the Dinosaur Catchment, suspended sediment typically forms >96 % of the total load, but seldom exceeds medium concentration (BEVERAGE & CULBERTSON 1964) and is well below the concentrations measured in sheet and rillwash during rainfall simulation experiments on hillslope microcatchments (BRYAN & HODGES 1984). Although these measurements combine suspended and bedload transport, the discrepancy nevertheless indicates substantial storage on hillslopes. Local bedrocks produce limited coarse debris and the largest fragments, derived from siderites, rarely exceed 25 mm in diameter. Most bedload is derived from sandstones, siltstones and silty micropediments with an average D_{50} of 0.11 mm. Measurements indicate that bedload typically forms <2% of the total load but observations of residual bedload suggest that the sampling procedures used may give conservative estimates.

In low intensity storms supply of sediment from hillslopes may exceed channel transport capacity, and produce some aggradation. The D_{50} coincides with the most easily entrained size fraction, with a threshold entrainment velocity of about 0.25 m s^{-1} which is achieved whenever channel flow depths reach 0.15 m. This means that virtually all bed material is mobile in medium or high intensity storms and bedload transport rates are limited only by hillslope supply rates. In these circumstances aggradation is negligible except in the last stages of recession when tunnel flow may contribute significant amounts of sediment but comparatively little discharge. The state of channel aggradation clearly depends on the relative frequency of storms or winter thaws of different intensity, both of which can very greatly in this area from year to year.

Regolith materials in the Katiorin Catchment are more variable in both erodibility and runoff response than those in the Dinosaur Catchment, and are also more clearly segregated within the catchment. Suspended sediment concentrations are very sensitive to the location of runoff, and while concentrations still fall in the medium range, sediment yields are significantly lower than in the Dinosaur Catchment due to the lower efficiency of runoff generation. Conversely, bedload apparently forms a much higher proportion of the total load in the Katiorin Catchment. In part this is due to the unusual contribution of tunnel-

Katiorin	Annual denudation	1.69 mm	(Minimum without large storms)
	Small storms	0.64 mm	
	Moderate storms	1.05 mm	
	Large storms	?	
Dinosaur	Annual Denudation	3.0–3.5 mm	(Long-term average)
	Small storms	0.84 mm	
	Moderate-Large storms	1.24 mm	
	Winter thaws	0.5–1.1 mm	

Tab. 5: *Summary of the influence of storms on denudation.*

induced clay slope failures and the production of armoured mudballs, but it may also reflect the hydraulic efficiency of a wider, smoother channel and projection of large particles into, or through, the flow. While, because of the nature of the lithology, the boundary between suspended and bedload transport in the Dinosaur Catchment is not well-defined, in the Katiorin Catchment it is much more distinct, as bedload has a D_{50} of 1.8 mm.

Most regolith materials in the Dinosaur Catchment are saline or sodic, but although electrical conductivities are high, the solute load seldom reaches 2% of the total load. The Katiorin clays are even more extreme in character than those in Dinosaur and potentially yield significantly larger amounts of solutes. The effects of highly saline clays and mudstones (tab.2) are counterbalanced by extensive regolith with low solute production potential in the upper catchment, so that solutes form roughly the same percentage of the total storm yield as in the Dinosaur Catchment (tab.4). Although this proportion is small the chemistry of the smectitic claystones and mudstones is important in both areas, strongly influencing shrink-swell capacities and therefore contributing to the instability of clay slopes, to desiccation cracking and to tunnel development which is well-advanced in both catchments.

The salient features of the storms described from each catchment are shown in tab.4. A number of assumptions have been used, as noted, to supply missing data to permit reasonable estimates of the proportional composition of storm yield and denudation rates. In tab.5 this information has been combined with data on the recurrence frequency of storms of different magnitude to provide estimates of overall annual denudation rates. Estimates for the Katiorin Catchment are certainly conservative as no large storms are included within the storm record. In view of this the estimated annual denudation rate of 1.69 mm agrees well with previous estimates of 2.2 and 2.3 mm yr^{-1} for the Chemeron and Endao catchments (BP-SAAP 1983). In the case of the Dinosaur Catchment, the estimated total annual rate of 2.08 mm yr^{-1} agrees well with storm-based estimates from other years (BRYAN & CAMPBELL 1986), and with long-term rates assessed by various methods (CAMPBELL 1974). While data for winter processes are too scant to allow independent estimates of their contribution to annual denudation, comparison of estimates for summer storms with long-term data suggests that the winter contribution could range between 15 and

30% depending on snow conditions.

Acknowledgement

Research was supported by individual operating grants from the Natural Sciences and Engineering Research Council, Canada, to R. Bryan and I. Campbell, and a fellowship from the International Development Research Centre to R. Sutherland. Research permits were issued by the governments of Kenya and Alberta. The assistance of Park Ranger R. Benoit and staff in Dinosaur Provincial Park, colleagues in Kenya and Alberta, and the Physical Geography Laboratory at the University of Amsterdam where clay mineral and regolith chemistry were analyzed, is gratefully acknowledged. The comments of two reviewers were helpful in revising the manuscript.

References

ANDERSON, D. (1981): Grazing, goats and government: ecological crises and colonial policy in Baringo. 1981–1939. Unpublished staff seminar, Department of History, University of Nairobi.

ATMOSPHERIC ENVIRONMENT SERVICE (1978): Rainfall intensity, duration and frequency values for Brooks horticultural station. Alberta: Hydrometeorology Division.

BOWYER-BOWER, T.A.S. & BRYAN, R.B. (1986): Rill initiation: concepts and experimental evaluation of badland slopes. Zeitschrift für Geomorphologie, Supp. Bd. **60**, 161–175.

BEVERAGE, J.P. & CULBERTSON, J.K. (1964): Hyperconcentrations of suspended sediment. Proceedings of the American Society of Civil Engineers, Journal of the Hydraulics Division, HY **6, 90**, 117–128.

BARINGO PILOT SEMI-ARID PROJECT (1983): Water development planning note: estimated erosion rates in the Chemeron and Endao catchments. Unpublished.

BARINGO PILOT SEMI-ARID PROJECT (1984): Interim Report 1984. Chapters 2 and 3. Unpublished.

BELL, H.S. (1940): Armoured mudballs: their origin, properties and role in sedimentation. Journal of Geology, **48**, 1–31.

BRYAN, R.B., YAIR, A. & HODGES, W.K. (1978): Factors controlling the initiation of runoff and piping in Dinosaur Provincial Park badlands, Alberta, Canada. Zeitschrift für Geomorphologie, Supp. Bd. **29**, 151–168.

BRYAN, R.B. & CAMPBELL, I.A. (1982): Surface flow and erosional processes in semi-arid mesoscale channels and drainage basins. International Association of Hydrological Sciences, **137**, 123–133.

BRYAN, R.B. & HODGES, W.K. (1984): Runoff and sediment transport dynamics in Canadian badland microcatchments. In: Burt, T.P. & Walling, D.E. (Eds.), Catchment Expriments in Fluvial Geomorphology. Geobooks (Norwich), 115–132.

BRYAN, R.B., IMESON, C.A. & CAMPBELL, I.A. (1984): Solute release and sediment entrainment on microcatchments in the Dinosaur Park badlands, Alberta, Canada. Journal of Hydrology, **71**,79–106.

BRYAN, R.B. & HARVEY, L.E. (1985): Observations on the geomorphic significance of tunnel erosion in a semi-arid ephemeral drainage system. Geografiska Annaler, **67**, 257–273.

BRYAN, R.B. & CAMPBELL, I.A. (1986): Runoff and sediment discharge in a semi-arid ephemeral drainage basin. Zeitschrift für Geomorphologie, Supp. Bd. **58**, 121–143.

BRYAN, R.B., CAMPBELL, I.A. & YAIR, A. (1987): Postglacial geomorphic development of the Dinosaur Provincial Park badlands, Alberta. Canadian Journal of Earth Sciences, **24**, 135–146.

CAMPBELL, I.A. (1974): Measurement of erosion on badland surfaces. Zeitschrift für Geomorphologie, Supp. Bd. **21**, 122–137.

CAMPBELL, I.A. (1981): Spatial and temporal variations in erosion measurements. International Association of Hydrological Sciences Publication, **133**, 447–456.

CHAPMAN, G.R., LIPPARD, S.J. & MARTYN, J.E. (1978): The stratigraphy and structure of the Kamasia Range, Kenya Rift Valley. Journal of the Geological Society, London, **135**, 265–281.

FROSTICK, L.E., REID, I. & LAYMAN, J.T. (1983): Changing size distribution of suspended sediment in arid-zone flash floods. Special Publication of the International Association of Sedimentologists, **6**, 97–106.

GARDINER, V. (1975): Drainage basin morphometry. British Geomorphological Research Group Technical Bulletin **14**.

HARTY, K.M. (1984): The geomorphic role of snow in a badland watershed. Unpublished M.Sc. thesis, University of Alberta.

HODGES, W.K. & BRYAN, R.B. (1982): The influence of material behaviour on runoff initiation in the Dinosaur badlands. In: BRYAN, R.B. & YAIR, A. (Eds.), Badland Geomorphology and Piping. Geobooks (Norwich), 13–46.

HODGES, W.K. (1984): Experimental study of hydrogeomorphological processes in Dinosaur Badlands, Alberta, Canada. Unpublished Ph.D. thesis, University of Toronto.

HONSAKER, J.L., CAMPBELL, I.A. & BRYAN, R.B. (1984): Remote semiautomatic instrumentation for intermittent streamflow measurement and suspended sediment sampling. Canadian Journal of Civil Engineering, **11**, 993–996.

KING, B.C. (1970): Vulcanicity and rift tectonics in East Africa. In: CLIFFORD, T.N. & GOSS, I.G. (Eds.), African Magmatism and Tectonics. Oliver & Boyd (Edinburgh), 263–283.

KOSTER, E.H. (1984): Sedimentology of a foreland coastal plain: Upper Cretaceous Judith River Formation at Dinosaur Provincial Park. Field trip Guidebook, Canadian Society of Petroleum Geologists, Alberta Research Council.

LEOPOLD, L.B. & EMMETT, W.W. (1981): Some observations on the movement of cobbles on a streambed. International Association of Hydrological Sciences, **133**, 49–59.

MINISTRY OF WATER DEVELOPMENT (1978): Rainfall frequency atlas of Kenya for durations from 10 minutes to 24 h. Master planning section, Water Department (Nairobi). 116 p.

POESEN, J. (1987): Transport of rock fragments by rill flow: a field study. In: BRYAN, R.B. (Ed.), Rill Erosion: Processes and Significance. CATENA Supp. Bd. **8**, 35–54.

ROBERTS, M. (1985): Fuel and Fodder Project, Baringo District. Progress Report, November 1983–March 1985. Unpublished.

RODGERS, J.A., DENNETT, M.D. & STERN, R.D.A (1982): Rainfall at Perkerra, Kenya. Tropical Agricultural Meteorology Group, University of Reading, Report **4**.

ROWNTREE, K.M. (1986): Annual, seasonal and storm rainfall on the Njemps Flats, Baringo, Kenya. Baringo Erosion Survey, Working Paper 3. Unpublished.

SCHICK, A.P. (1977): A tentative sediment budget for an extremely arid watershed in the southern Negev. In: DOEHRING, D.O. (Ed.), Geomorphology in Arid Regions. Proceedings of the Eighth Annual Geomorphology Symposium (Binghampton), 139–163.

STOCKING, M.A. & ELWELL, W.A. (1976): Rainfall erosivity over Rhodesia. Transactions of the Institute of British Geographers, **1**, 231–235.

SUTHERLAND, R.A. (1983): Mechanical and chemical denudation in a semiarid badland environment, Dinosaur World Heritage Park, Alberta, Canada. Unpublished M.Sc. thesis, University of Toronto.

TALLON, P.W.J. (1978): Geological setting of the hominid fossils and Acheulian artifacts from the Kapthurin Formation, Baringo District, Kenya. In: BISHOP, W.W. (Ed.), Geological Background to Fossil Man. Geological Society (London), 361–373.

YAIR, A. & LAVEE, H. (1974): Areal contribution to runoff on scree slopes in an extreme arid environment; a simulated rainstorm experiment. Zeitschrift für geomorphologie, Supp. Bd. **21**, 106–121.

YAIR, A., SHARON, D. & LAVEE, H. (1980): Trends in runoff and erosion processes over an arid limestone hillside, northern Negev, Israel. Hydrological Sciences Bulletin, **25**, 243–255.

Addresses of authors:
R.B. Bryan and R.A. Sutherland
Department of Geography
University of Toronto (Scarborough Campus)
1265, Military Trail, Scarborough
Ontario M1C 1A4
Canada
I.A. Campbell
Department of Geography
University of Alberta
Edmonton, Alberta, T6G 2H4
Canada

R. B. Bryan (Editor)

RILL EROSION

Processes and Significance

CATENA SUPPLEMENT 8

192 pages / hardcover / price DM 149,— / US $ 88.—

Special rate for subscriptions until

December 15, 1987: DM 119,— / US $ 70.40

Date of publication: July 15, 1987   ORDER NO. 499/00 107

ISSN 0722-0723/ISBN 3-923381-07-7

CONTENTS

CATENA SUPPLEMENT 13 p. 37–47 Braunschweig 1988

MEDIUM-TERM EROSION RATES IN A SMALL SCARCELY VEGETATED CATCHMENT IN THE PYRENEES

N. **Clotet-Perarnau**, F. **Gallart** & C. **Balasch**, Barcelona

Summary

Badland areas on Cretaceous mudrocks are the chief sources of sediment in the high Llobregat basin. An earlier one-year study carried out in 1982, showed annual erosion rates of 14 kg/m^2 from weights of sediment collected from a catchment of 37.5 m^2 in area, and 55 kg/m^2 from exhumed depth of iron pins. Three years later, the rate measured on iron pins showed only 23 kg/m^2 per year, much nearer to the value yielded by weighing the sediment. Such values are however not too reliable as long-term estimates because of the occurrence of an extreme rainfall in November, after a relatively rainy August.

Near to the small village of Vallcebre (High Llobregat basin), a catchment of 0.031 km^2 was dammed 40 years ago by a rock slide. The volume of sediment trapped by this dam has provided us with an excellent tool to check the medium-term validity of our measures. The heterogenity of the hillslopes of this small catchment is a handicap in attempting to relate the volume of sediment to a depth eroded; although, if an insignificant value is assumed as supplied from wooded areas, the erosion rates are of the same magnitude as those measured by weighing the sediment in 1982.

The reliability of the rates obtained in 1982 suggests to us that the gross erosion rates of these badlands are controlled by the rate of weathering rather than rainfall events.

Resumen

Las áreas de badlands en las arcillas del Cretácico son las pricipales fuentes de sedimento de la cabecera del Llobregat. Mediciones llevadas a cabo a lo largo del año 1982 del peso del sedimento evacuado de una cuenca de 37.5 m^2 dieron tasas de 14 kg m^{-2}, y alrededor de 55 kg m^{-2} en la medición de la denudación llevada a cabo con clavos. Tres años más tarde la tasa obtenida a partir de la medición de los clavos fué de 23 kg m^{-2} a^{-1}, valor mucho más cercano al encontrado con la valoración hecha a patir del peso del sedimento. Estas cifras no son sin embargo muy fiables como estimaciones a largo plazo por la ocurrencia, durante el período de medición, de una precipitación extraordinaria en el mes de noviembre, precisamente después de un mes de agosto relativamente lluvioso.

Cerca del pequeño pueblo de Vallcebre (Alto Llobregat) una cuenca de 0.031 km^2 de árez quedó obturada hace

ISSN 0722-0723
ISBN 3-923381-13-1

40 años a causa de un deslizamiento de rocas. El volumen del sedimento acumulado en este embalse durante estos años nos ha proporcionado una herramienta excelente para comprobar la fiabilidad a término medio de nuestras mediciones. Si bien la heterogeneidad de las vertientes de esta pequeña cuenca resulta un inconveniente para relacionar el volumen de sedimento con la produndidad de la capa erosionada, si atribuimos un valor insignificante al material procedente de las áreas de bosque, las tasas de erosión son entonces muy similares a las calculadas a partir del peso del sedimento movilizado en 1982.

La fiabilidad de las tasas obtenidas en 1982 nos confirma que la actividad de estos badlands está controlada por la tasa de meteorización más que por los acontecimientos lluviosos.

1 Introduction

Erosion rates in badland areas show a wide range of values and their assessment presents several kinds of problem. Worldwide average measured rates range between 0.45 mm/year (YAIR et al. 1982) and 20–30 mm/year (ALEXANDER 1982), both from within the Mediterranean region. Rates in continental high-altitude areas tend to be high because of frost weathering and mass movement during snowmelt period. Arid areas show the lowest rates as a consequence of the short duration of storms and the influence of long dry periods (BRYAN & YAIR 1982), but Montmorillonite-rich clays can however produce very active badland areas in Mediterranean climates because of the combination of running water and mass movement processes (ALEXANDER 1982).

Estimates of erosion rates depend not only on climatic and lithologic characteristics, but also on the techniques used and on the spatial and temporal scales involved. Spot short-term rates usually yield much higher figures than long-term estimates for whole drainage basins, reflecting the high variability of surface conditions at a detailed scale and the difficulty of assessing reliable estimates of sediment delivery from badland areas (YAIR et al. 1980).

In order to establish the significance of land degradation in some badland areas on sediment supply in the high Llobregat valley (Eastern Pyrenees), estimates of erosion rates were needed. During 1982 monitoring of spot erosion rates was carried out (CLOTET & GALLART 1986), which provided us with values ranging between 37 mm/year (erosion pins), 9.3 mm/year (sediment collected from a small catchment) and 7.3 mm/year (sediment collected from a small hillslope). A sediment budget based on the second of these rates showed us the great role of badland areas in the sediment supply to the drainage net. This figure was, however, of doubtful reliability as a long-term estimate because of the occurrence of extreme rains during the monitoring period.

The aim of the present paper is to analyse the reliability of the former figures as medium-term estimates, and the role of weathering, erosion and transport as controls of sediment yield rates for changing spatial scales.

2 Characteristics of the Study Area

The Vallcebre basin lies in the southern belt of the eastern Pyrenees, in the

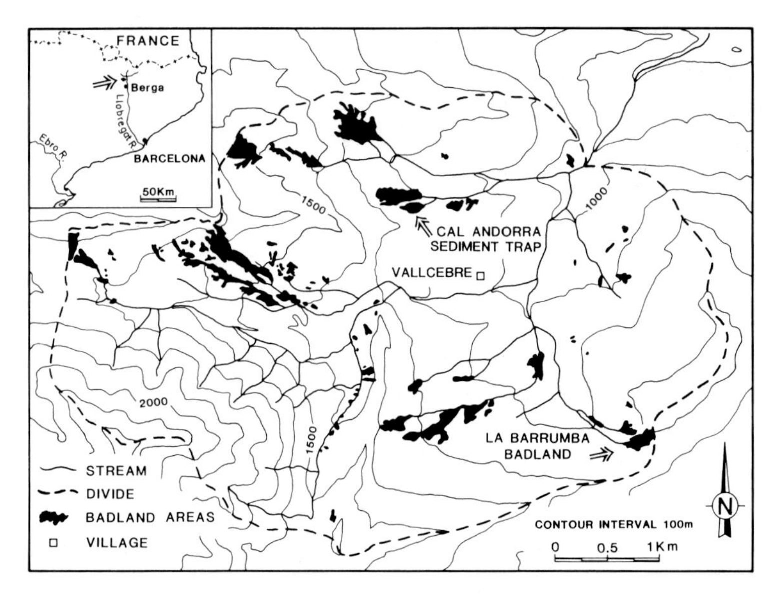

Fig. 1: *Location of the Vallcebre basin and general pattern of badland area.*

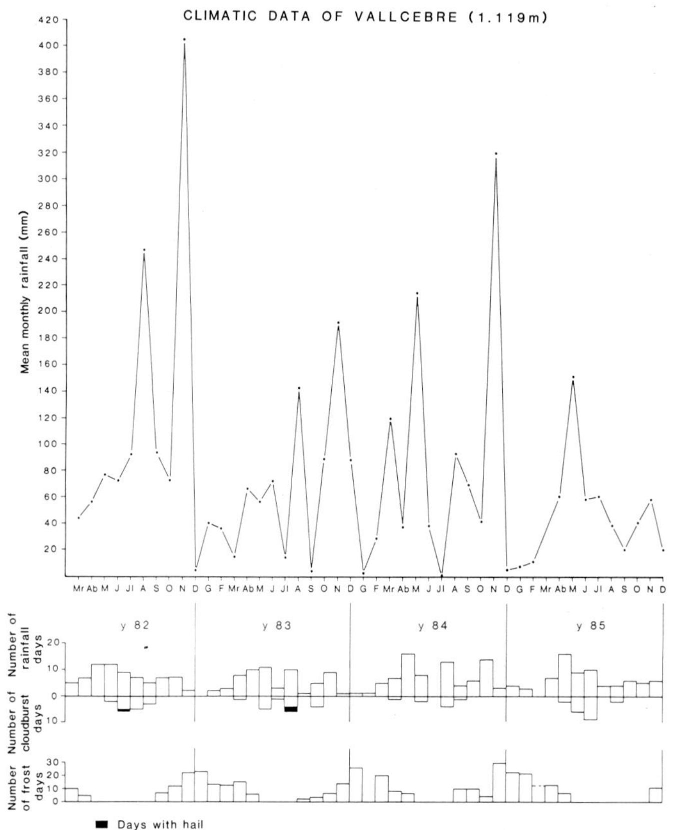

Fig. 2: *Rainfall data from the raingauge at Vallcebre.*

headwaters of the Llobregat river, at elevation between 900 and 2300 m a.s.l. (fig.1). The highest part of the catchment consists of limestones and sandstones of late Cretaceous age, folded in a dome structure. The rest of the area, much more extensive, is in continental montmorillonite-rich mudstones of Garumnian facies (Cretaceous-Paleocene boundary) with several beds and banks of lacustrine limestones and calcareous sandstones. Both lithologic units belong to the Pedraforca Nappe, which has been interpreted as overthrusted by more than 20 km from the north.

The mean annual rainfall at Vallcebre (1119 m a.s.l.) is about 940 mm, although the record started only in 1982 and the long-term value is expected to be somewhat higher. Rainfall amounts show a very irregular distribution through the year. Two main peaks correspond to spring and autumn, and another peak, somewhat less important, but with high intensities occurs in summer, corresponding to convective storms, sometimes accompanied by hail (fig.2). Snow falls on about 22 days/year, usually between November and April, but sometimes as late as May. The mean annual temperature at Figols 10 km to the south, is 11.2°C. Frost occurs on about 100 days per year.

The climate, the calcareous character of most of the rocks and the human activity are the main controls of the vegetation cover. Below 1600 m of altitude the forests of *Pinus sylvestris* very often replace the *Buxo-Quercetum pubescentis* association, which is present only in some remnants. Forests of *Fagus sylvatica* cover come of the north-facing slopes. Above 1600 m the association *Pulsatillo-pinetum uncinatae* is the dominant, with a grass layer and some groups of *Juniperus communis* (MASALLES & SEBASTIA 1985). Below 1300 m the great majority of gentle slopes are terreced for agricultural use, now mostly abandoned.

3 The Barrumba Badland: A One Year Monitoring Period

In order to determine erosion rates in the badland areas on Garumnian clays we selected a single badland area with an area of 0.02 km^2, and at an altitude of 1240 m (fig.3, photo 1, 2). This area was instrumented with very simple instruments (6 erosion pins and a sediment-collecting plastic bag on hillslopes, and a plastic bag which collected the sediment coming from a small catchment of 37.5 m^2), and was regularly visited during 1982 in order to maintain the instruments, to draw cross-sections on the main channels, and to take repeated photographs. Unfortunately no autographic rainfall data, but only cumulative rainfall totals between visits, are available for this period, the analysis of the rainfall events being based on the data from the raingauge of Vallcebre, 2 km from the study area.

The results of this study showed great seasonal differences in the geomorphic processes, revealed by both qualitative and quantitative data (CLOTET et al. 1983, CLOTET & GALLART 1986). During winter, frost action is able to soften the mudrock and a weathered cover a few centimeters deep appears at springtime. Snow melting or weak spring rains produce saturation of this weathered mantle and some flow scars and chutes are visible on slopes, while flowing materials accumulate in the main rills or on footslopes. Wind and sun-

Photo 1: *Intensely dissected mudrock slope in La Barrumba badland area.*

Photo 2: *Irregular slope in La Barrumba badland. Steeper slopes are cut in lime or gypsum-rich mudrock beds. On the right side, coarse debris fill an elementary channel and impede vertical erosion.*

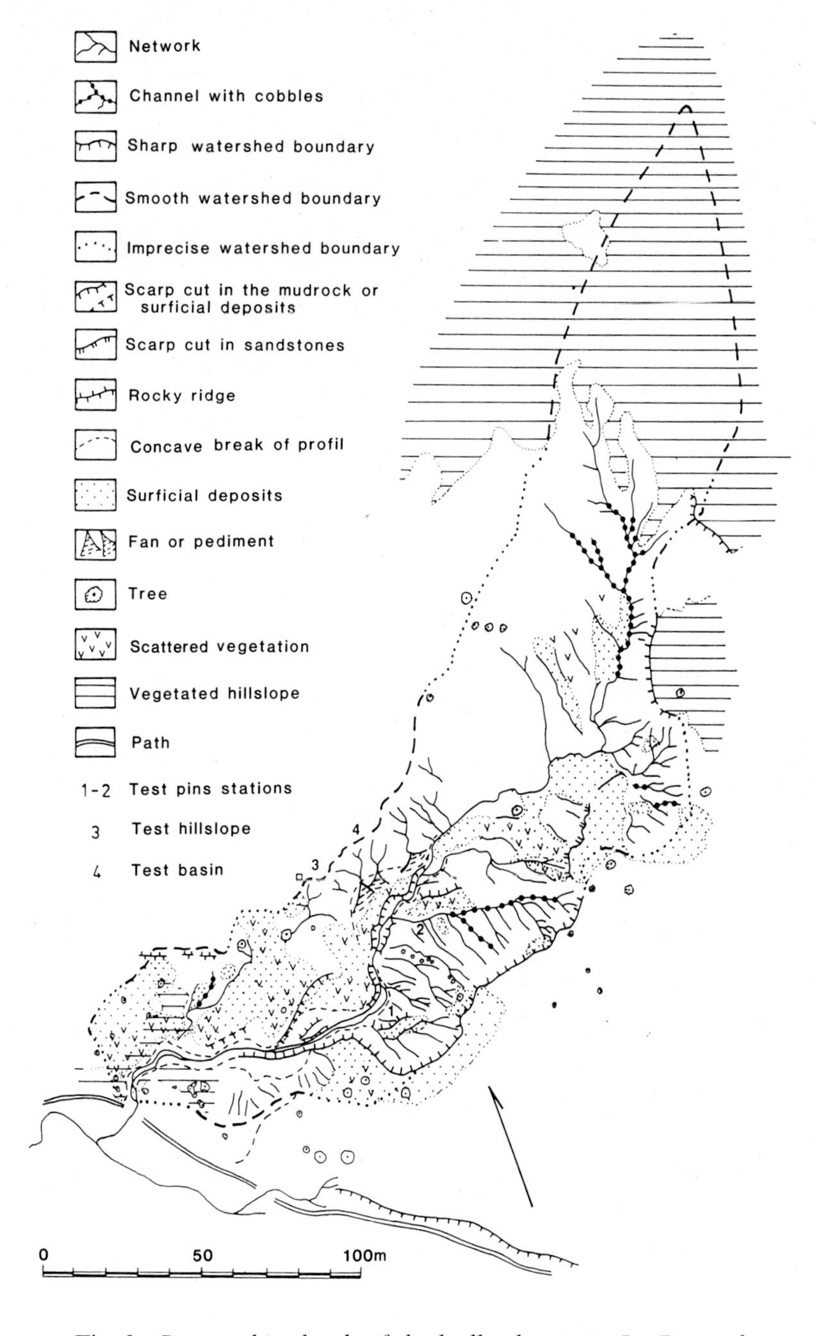

Fig. 3: *Geomorphic sketch of the badland area in La Barrumba.*

shine dessication produce small mud fragments which collapse and form small talus deposits on most of the foot-slopes. In summer, convective down-pours sometimes accompanied by hail, produce rapid erosion of most of the weathered mantle, and its deposition in the main channels. On 9th August 1982 we recorded an event which produced the half of the annual erosion on slopes, deposited a similar amount of sediment in the channels, and increased the relationship between erosion and rainfall for subsequent events. As we have no eye-witness to the event, we assumed that hail-fail could explain these facts, by rapid erosion without subsequent runoff, and by destruction of the crust which formerly protected the regolith against rainsplash erosion (see CLOTET & GALLART 1986). The sediment stored in the channels needs long-lasting runoff events during autumn or early winter to be flushed away, but some of the sediments deposited in fans or on pediments constitute semi-permanent stores.

Erosion rates were measured only on bare clay areas which appeared to be much more active than debris-covered areas, and occur on about 50% of the whole badland area. In spite of the good link between the main erosion events as registered by sediment traps and erosion pins, sediment trap data suggest an annual erosion rate of 14 kg/m^2 (9.3 mm of surface lowering) from a small 37.5 m^2 catchment and 11 kg/m^2 (7.3 mm) from an open hillslope, while erosion pin measurements suggest as much as 37 mm of surface lowering, equivalent to 55 kg/m^2. Three years later, the annual rate measured on iron pins was only 15.3 mm, or 23 kg/m^2, much nearer to the rate obtained by weighing the sediment. In order to avoid site disturbances we selected different areas for the two techniques, but it is obvious from our results that both kinds of measurements must be done in a manner that is directly comparable. However, erosion pins seem usually to yield higher rates than those yielded by sediment trapping techniques (YAIR, personal communication).

4 The Cal Andorrà Catchment: A 40-Years Old Sediment Trap

A planar slide of a limestone bed over clays dammed a 0.031 km^2 catchment near the village of Vallcebre, 3 km from La Barrumba and on the same kind of bedrock, in 1945 (photo 3, 4). The date of the mass movement was established from one of the inhabitants of Vallcebre who remembered the date because of the coincidence with a personal event. Moreover, the age of the oldest trees growing over the moved material yielded, by tree-ring dating, 32 years in 1986; 9 years is an acceptable delay between the movement and seedling growth.

The sediment trap was surveyed on April, 1985 with a planetable, and 21 bores were drilled with a manual boring device (see fig.4, bottom); the volume, calculated by drawing the hypsometric curve, yielded 2100 m^3. A supplementary continuous-sample drilling showed the whole volume was built up by 13 fining upwards graded units, that we assumed as separate filling events; their mean volume is 160 m^3 with a standard deviation of 42.3 (fitting with a normal distribution can be accepted on the basis of the Kolmogorov-Smirnov test).

Unfortunately, the hillslopes of this catchment are very heterogeneous and it is not possible to compute a sin-

Photo 3: *Rocky planar slide which dammed the Cal Andorrà catchment in 1945. View from the north.*

Photo 4: *Filling plain in the sediment trap. View from the east.*

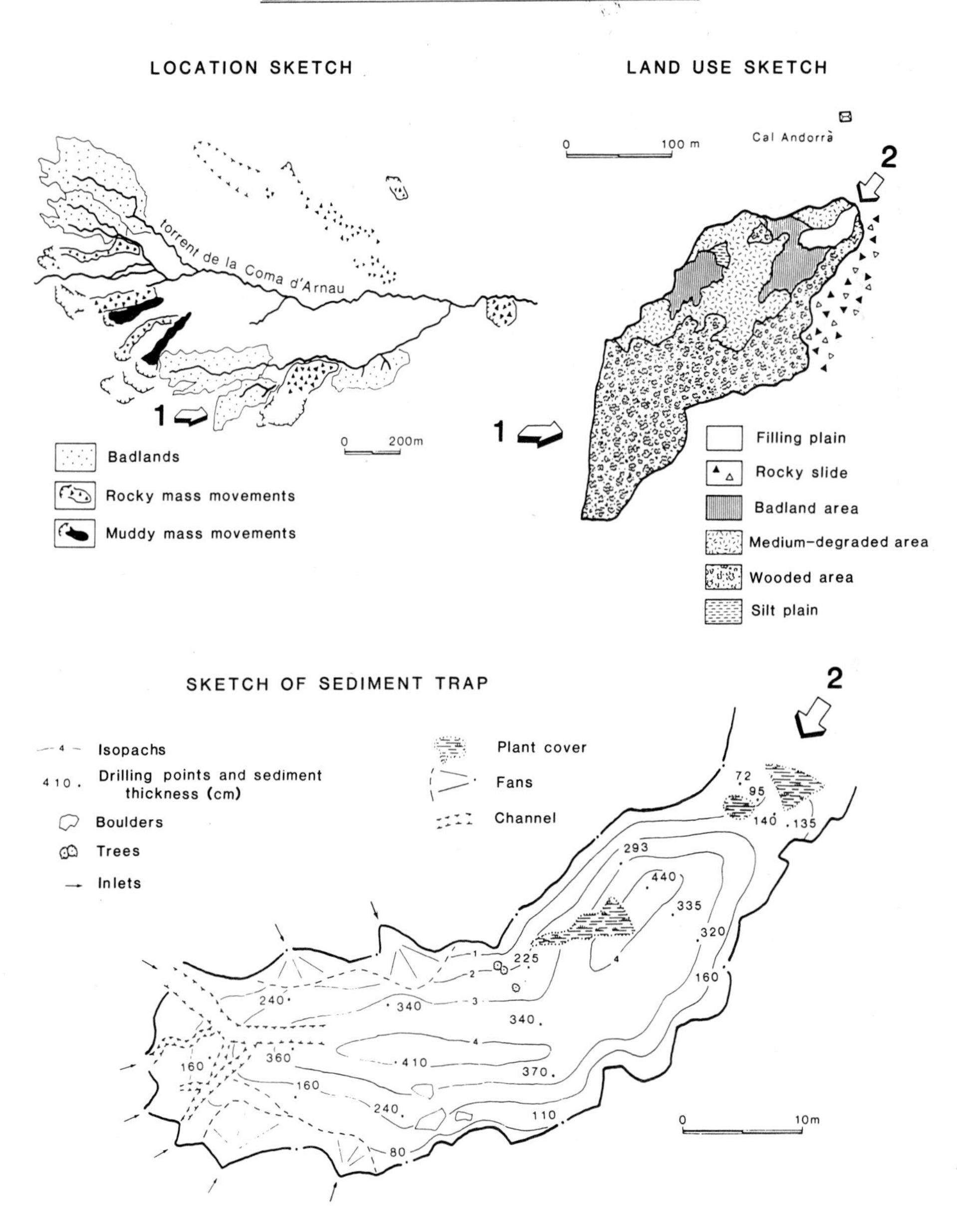

Fig. 4: *The Cal Andorrà sediment trap. Location, vegetal cover of the catchment, and topography of the sediment trap.*

gle erosion rate. Badland areas cover 12% of the area, medium-degraded areas (patches of grass with scattered trees alternating with bare soil patches where erosion is clearly active) 26.2%, wooded stable areas (most of them on limestones) 58%, and depositional plains 3.5% (fig.4). The mean erosion rate, taking 0.83 as the relationship between the bulk density of the sediments in the trap and that of the weakly weathered mudrock, would be 1.4 mm/year, but, owing to the heterogenity of the catchment, it is of limited value.

If the badland areas are considered as the only sediment source, we can estimate an upper erosion rate of 12 mm per year. A lower estimate can be made by assuming that the stable wooded areas supply an insignificant amount of sediment, but the sediment comes equally from badland and medium-degraded areas. This estimate would be 3.8 mm per year. The values obtained by erosion pins in la Barrumba fall outside this range, while the figures obtained by sediment trapping are are within it, somewhat nearer to the upper limit but for a 40-years record and for a drainage area a thousand-fold larger. If 9.3 mm per year, the figure obtained by weighing the sediment yielded by the small catchment in La Barrumba, is taken as the best estimate of erosion rates in badland areas, the resultant erosion rate for the medium-degraded areas is 1.2 mm per year.

From the number of the filling events found in the drilling cores, their recurrence interval is about three years. If we accept the main pattern of the processes abserved in La Barrumba, these filling events seem therefore to be less related to the weathering-erosion seasonal couples than to runoff events which could convey the sediments stored in footslopes and channels and flush them out.

5 Discussion and Conclusions

On the basis of the short-term small sediment trap data (La Barrumba) and medium term sedimentation data (Cal Andorrà), erosion rates as high as 9 mm/year appear to be representative of badland areas on Garumnian mudstones in this mountainous Mediterranean region. This high erosion rate is the result of frost weathering during winter and strong erosion by summer downpours sometimes accompanied by hail. Mass movement processes are active during snowmelt by mudsliding and during dry periods by shrinking and falling of dry loose fragments, but their role is of secondary importance.

Short-term measurements at La Barrumba were carried out during 1982, a year with a rainy summer and an extreme continuous rainfall event in November, but the rates obtained there fall within the interval we can accept for the 40-years old record at Cal Andorrà. Furthermore, we observed at La Barrumba that most of the regolith was eroded from the hillslopes during the first summer downpour, and subsequent rains produced much less sediment. It is therefore suggested that erosion processes and rates are controlled by mudstone weathering rates, while rainfall events control long-term sediment transport, which shows a rather irregular temporal pattern.

The high erosion rates obtained raise the question of the origin of these badland areas. The median of the sizes of badland areas in the Vallcebre basin is 2300 m^2 most of them being therefore of recent origin. Mass movements triggered by extreme rains produce unprotected soil or mudrock exposures able to develop gullies and badland areas; the new badland-like areas which have recently appeared were indeed produced by shallow slides as a consequence of the extreme rain of November 1982, and there are several reported incidences of badland areas triggered by such phenomena (TRICART 1974, HARVEY 1986). Another origin can be attached to human disturbance. Old deforestation for agricultural and stockbreeding land use and, more recently for mining, can be a trigger for gullies or badland areas; the lack of trees can moreover increase the occurrence of shallow mass movements on relatively gentle slopes during extreme rainfall events (GALLART & CLOTET, this volume).

Acknowledgement

We acknowledge the comments from Prof. A. Yair, and the field work help from J. Parga. We are also very grateful to Dr. A.M. Harvey and the anonimous referees from CATENA for their critical review of the amnuscript and English style improvements.

References

ALEXANDER, D. (1982): Difference between "calanchi" and "biancane" badlands in Italy. In: BRYAN, R. & YAIR, A. (Eds.), Badland Geomorphology and Piping. Geo-Books, Norwich, 71–87.

BRYAN, R. & YAIR, A. (1982): Perspectives on studies of badland geomorphology. In: BRYAN, R. & YAIR, A. (Eds.), Badland geomorphology and Piping. Geo-Books, Norwich, 1–12.

CLOTET, N., GALLART, F. & CALVET, J. (1983): Estudio de la dinámica de un sector de badlands en Vallcebre (Prepirineo catalán). Actas de la II Reunión del Grupo Español de Geología Ambiental y Ordenación del Territoria, Lleida, 4.20–4.38.

CLOTET, N. & GALLART, F. (1986): Sediment yield in a mountainous basin under high Mediterranean climate. Zeitschrift für Geomorphologie Supp. Bd. **60**, 205–216.

GALLART, F. & CLOTET, N. (this volume): Some aspects of the geomorphic processes triggered by an extreme rainfall event: the November 1982 flood in the Eastern Pyrenees.

HARVEY, A.M. (1986): Geomorphic effects of a 100 year storm in the Howgill Fells, Northwest England. Zeitschrift für Geomorphologie N.F. **30**, 71–91.

MASALLES, R. & SEBASTIA, T. (1985): Descripció de la vegetació i mesures de revegetació per a la fi de l'extracció de minerals a les parcelles de Coll Pradell i el Torrent de les Llobateres (Serra d'Ensija). E.N.H.E.R. unpublished report.

TRICART, J. (1974): Phénomènes demesurés et régime permanent dans des bassins montagnards. Revue de Géomorphologie Dynamique **23**, 99–114.

YAIR, A., BRYAN, R.B., LAVEE, H. & ADAR, E. (1980): Runoff and erosion processes and rates in the Zin Valley badlands, northern Negev, Israel. Earth Surface Processes **5**, 205–225.

Address of authors:
N. Clotet-Perarnau, F. Gallart, C. Balasch,
Institut Jaume Almera
Ap 30102, 08028
Barcelona, Spain

F. Ahnert, H. Rohdenburg & A. Semmel:

BEITRÄGE ZUR GEOMORPHOLOGIE DER TROPEN (OSTAFRIKA, BRASILIEN, ZENTRAL- UND WESTAFRIKA) CONTRIBUTIONS TO TROPICAL GEOMORPHOLOGY

CATENA SUPPLEMENT 2, 1982
Price: DM 120,–
ISSN 0722–0723 / ISBN 3–923381–01–8

F. AHNERT
UNTERSUCHUNGEN ÜBER DAS MORPHOKLIMA
UND DIE MORPHOLOGIE DES
INSELBERGGEBIETES VON MACHAKOS, KENIA

(INVESTIGATIONS ON THE MORPHOCLIMATE
AND ON THE MORPHOLOGY OF THE
INSELBERG REGION OF MACHAKOS, KENIA)

S. 1–72

H. ROHDENBURG
GEOMORPHOLOGISCH–BODENSTRATIGRAPHISCHER
VERGLEICH ZWISCHEN DEM
NORDOSTBRASILIANISCHEN TROCKENGEBIET
UND IMMERFEUCHT–TROPISCHEN GEBIETEN
SÜDBRASILIENS

MIT AUSFÜHRUNGEN ZUM PROBLEMKREIS DER
PEDIPLAIN–PEDIMENT–TERRASSENTREPPEN

S. 73–122

A. SEMMEL
CATENEN DER FEUCHTEN TROPEN
UND FRAGEN IHRER GEOMORPHOLOGISCHEN
DEUTUNG

S. 123–140

CATENA SUPPLEMENT 13 p. 49–60 Braunschweig 1988

PIPING IN BADLAND AREAS OF THE MIDDLE EBRO BASIN, SPAIN

M. **Gutiérrez**, G. **Benito**, Zaragoza
J. **Rodríguez**, Huelva

Summary

Piping processes are studied in recent alluvial fans and on slopes. In both there are an extensive networks subsurface ducts which, on collapse, generate gully networks. The piping in alluvial fans develops as a consequence of intense surface microcracking and also because of the high degree of alkalinity and material dispersion. On the slopes the ducts originate where there is contact with the Miocene bedrock, and due to the high level of porosity of the detritus, give rise to large pseudo-dolines on the surface.

Resumen

Se estudian procesos de piping en abanicos aluviales holocenos y en depósitos de vertiente. En ambos se reconoce una extensa red de conductos subsuperficiales que por evolución a favor de colapsos van a generar una red de gullies. El piping en los abanicos aluviales se desarrolla como consecuencia de un intenso microagrietamiento superficial y también por la elevada alcalinidad y dispersión de los materiales. En el área de vertientes los conductos se originan en el contacto con el sustrato mioceno, debido a la elevada porosidad de los detritos, dando lugar en superficie a pseudo-dolinas de gran tamaño.

ISSN 0722-0723
ISBN 3-923381-13-1

3-923381-13-1/88/5011851/US$ 2.00 + 0.25

1 Introduction

The piping phenomenon is relatively common in different climatic environments and for several decades there has been an extensive bibliography on this subject (see JONES 1981). Over the last few years research has tended towards the study of the hydrology of subsurface networks generated by piping processes (GILMAN & NEWSON 1980, JONES 1982), morphometric analysis of the pipes and pipe networks (JONES 1981), and also chemical-physical studies to establish the piping genesis (CROUCH et al. 1986, HEEDE 1971). Another line of research has concentrated on the relationship between piping and badlands or gully systems (BRYAN & YAIR 1982).

A large number of piping studies are carried out in humid areas, where the ducts are preferentially situated in relation to textural variations in the soil profile. In arid and semi-arid morphoclimatic systems, piping processes appear to be linked to the presence of expansive clays and to high values of dispersion and exchangeable sodium percentages (HEEDE 1971). Furthermore, as

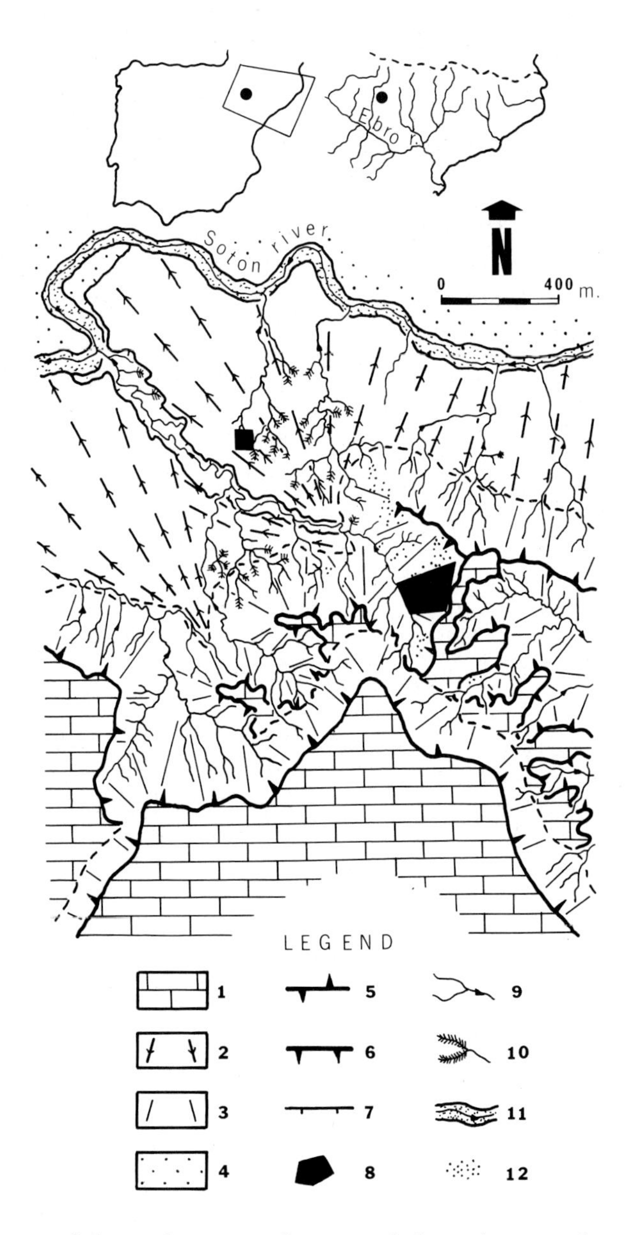

Fig. 1: *Situation of the study area and geomorphological map, 2 km to the south of Lupiñén (province Huesca).*

1 = Limestone; 2 = Alluvial fans; 3 = Coarse talus; 4 = Terrace deposits; 5 = Ridges; 6 = Structurally controlled sharp break of slope; 7 = Sharp break of slope in recent deposits; 8 = Study areas; 9 = Drainage network; 10 = Badlands; 11 = Alluvium; 12 = Pseudodolines

has already been stated, the pipes are found above all in badland areas and often at the head of drainage networks (LEOPOLD et al. 1964).

2 Study Area

This paper deals with piping in the middle Ebro Basin, northern Spain. The field site (fig.1) is about 15 km south of the Pyrenean 'Sierras Exteriores' and 2 km south of Lupiñén (Province of Huesca). The climate is continental mediterranean; the temperatures in Ardisa (14 km to the west) vary between 34°C average maximum temperature for July and 0°C average minimum temperature for January. The mean annual rainfall in Lupiñén is 650 mm, most of which is storm rainfall; some 15 storms per year occur between May and October, each with an average of 30 mm of rain. In the last 20 years the heaviest storm rainfall was that of 101 mm, which fell on October 22nd, 1982.

The bedrock comprises Miocene shales with thin sandstones and limestones. The uppermost part of the formation is strongly calcareous and dips gently southwards, forming a well developed cuesta with stepped profiles on its front, partially covered by a considerable amount of debris (up to 15 m). These slopes are dissected by a network of rills and gullies which, in the lower, more gently sloping parts, have deposited extensive coalescent alluvial fans. These have buried the River Sotón's 3–5 m terrace (fig.1). There are the two study areas: one situated on the recent alluvial fan deposits and the other on the debris slopes (fig.1).

3 Piping in Alluvial Fans

The deposits that make up the alluvial fan have a thickness of 5 m and are composed of fine detritus within which there are occasional intercalated channels of cemented gravles of up to 20 cm in thickness. These deposits are dissected by a network of rills and gullies, linked to numerous pipes.

We have mapped this micromorphology over an area of some 1,150 m^2, using a 5 m mesh (fig.2). As a whole, the network is discontinuous dentritic and the major water courses present a meandering pattern, both above and below the surface. A very common feature here are inlets (vertical pipes) which collect the water from a small network of rills (photo 1) and, on occasions, reemerge on the surface through outlets. These inlets can in time be replaced, towards the head of the network, by other inlets, the points of water loss thereby moving headwards and leaving inactive inlets, downstream. It can be clearly seen that the network heads are comprised of small rills or rather semicircular walls (square C′E), corresponding in some cases to collapsed pipes. The gully walls exposed by incision reveal the remains of a subsurface network perched at up to three different levels and at times going in different directions. This indicates temporal variation in the channel's route, with a morphology of abandoned meanders (E′K) and lateral displacements (band E′), which give rise to overhanging channel walls. The subsurface network is at times inferred from vertical pipes (C′N, D′N), gentle depressions with circular cracks and collapses that are often found to have pipes within them (A′K, A′J).

For a statistical analysis 688 pipes

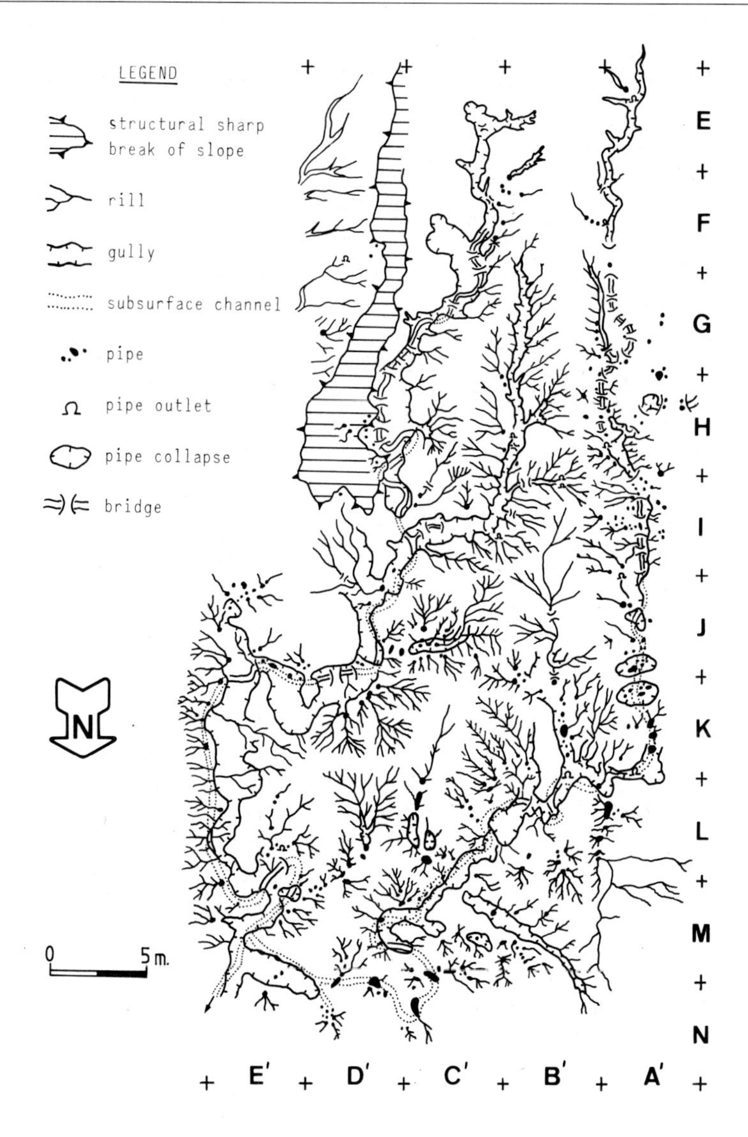

Fig. 2: *Detailed geomorphological map of the badlands area and pipes developed on the recent alluvial fan.*

have been measured; 51.5% are vertical, 22.5% inclined and 16% horizontal. The vertical and inclined pipes are normally smaller than 30 cm in diameter, although some do reach 60 cm. The most common size is between 2 and 6 cm and 75% of the pipes are smaller than 8 cm. The horizontal pipes also have a maximum diameter of 60 cm and there are 2 predominant sizes, 2–6 cm and 18–20 cm, and 75% of these pipes are under 20 cm in size. We have also differentiated between the morphology of those pipe sections in which different shapes

Photo 1: *Network of rills draining towards an inlet.*

are observed: circular, elliptical, lenticular, semicircular, key-hole, trapezoidal and rectangular. The predominant shape for vertical and inclined pipes is the elliptical (64.7%) and for horizontal ones the semi-circular and elliptical (33.5% each).

Fig.3a is a topographic map of the area shown in fig.2, which illustrates the low relief, and gentle slope, but showing steeper gradient in the lower areas of the main gullies. The density of pipes (fig.3b) indicates their adaptation to the principal gullies, with there being more pipes in the western network (above all at the head) and in the confluence areas. Lower densities occur near the interfluves. The distribution of vertical pipes (fig.3c) largely coincides with the total distribution, though their percentage decreases in the head area, and increases in the middle parts. On the other hand, for the horizontal pipes (fig.3d), no apparent relation in their distribution can be seen and they correspond to the outlets of the subsurface network. The maximum size of all the pipes (fig.3e and 3f) increases in general down the gully network.

From this analysis we can infer a close interrelation between the piping and gullying phenomena (photo 2). The pipes preferentially control the gully orientations, as is indicated by SLAYMAKER (1982) and by HARVEY (1982), and the gullies are generated as a result of progressive pipe collapse (DE PLOEY 1974) and, a rapid down-cutting (AGHASSY 1973). This last observation is made clear by the fact thet the sinuous gully network does not seem to have resulted from subaerial evolution, but by collapse from a sinuous subsurface drainage network (see locations C′M, D′M, C′N, D′N). From this we may deduce from this that the pipe network conditions the future

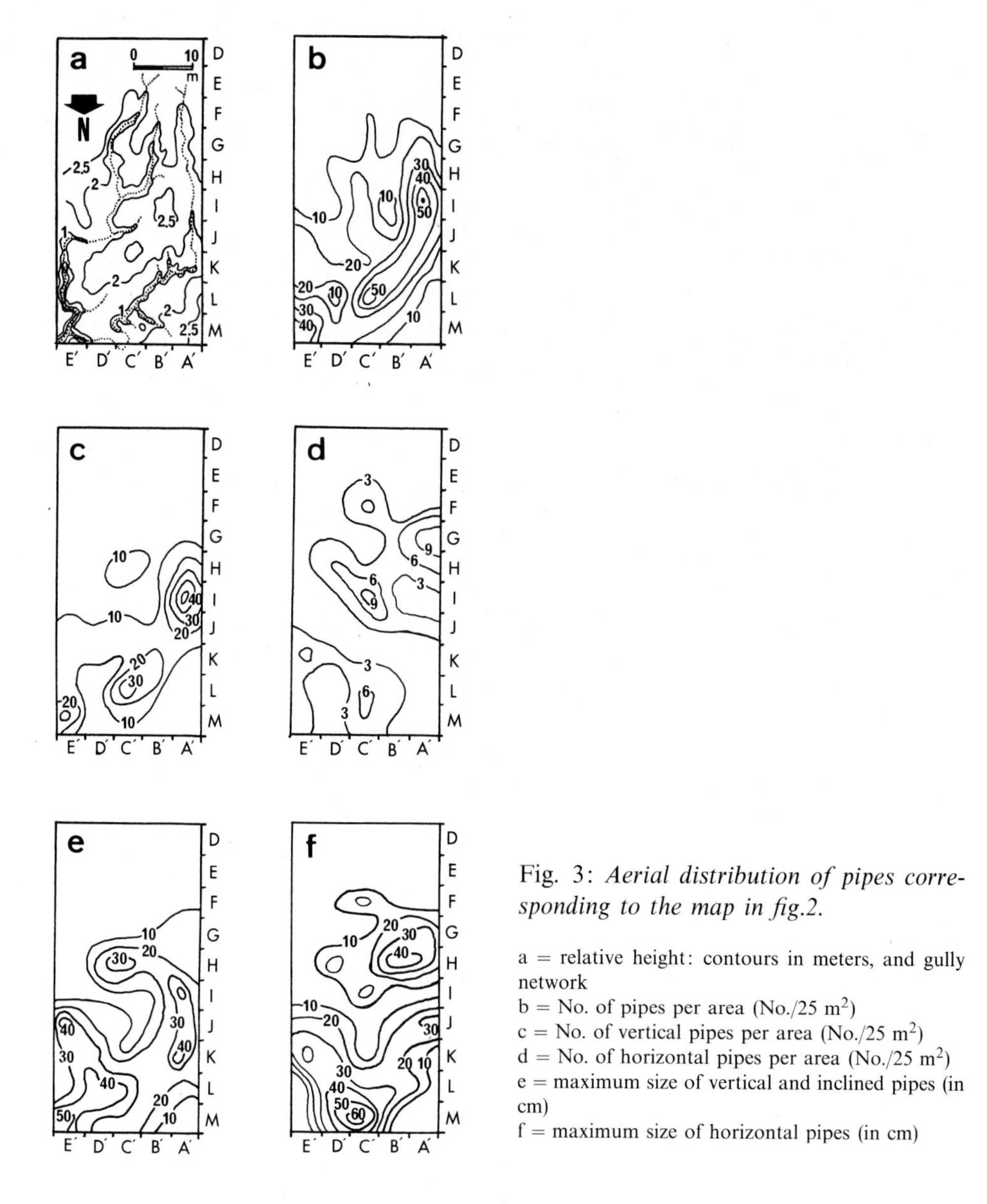

Fig. 3: *Aerial distribution of pipes corresponding to the map in fig.2.*

a = relative height: contours in meters, and gully network
b = No. of pipes per area (No./25 m^2)
c = No. of vertical pipes per area (No./25 m^2)
d = No. of horizontal pipes per area (No./25 m^2)
e = maximum size of vertical and inclined pipes (in cm)
f = maximum size of horizontal pipes (in cm)

gully network to a great extent. On the other hand, the embryonic gullies (C′J), generated by collapse and not connected superficially with the gully system, serve as rill collectors with small pipes, which in turn empty into an embryonic gully. Although the gully is initially generated by progressive collapse of the pipe network, it can, later on, be left suspended and inactive as a result of the appearance of a new pipe network.

These observations make it clear thet the gullying and piping processes are interdependent although in the early stages

Photo 2: *Different types of pipes and bridges associated with the development of a gully.*

the development of subsurface networks may be the initial cause of gullies in the head areas. Even so, there is some evidence that an incipient rill network could well be the collector of surface water which, via percolation, feeds to subsurface flow.

4 Piping on the Slopes

On the slopes at the foot of the limestone cuesta (fig.1) are a large number of closed depressions (photo 3). A sector of this slope, where their density is sufficiently representative has been mapped (fig.4). These slopes, with average gradients of 32%, are formed by debris of a highly variable thickness (up to 15 m), comprising angular blocks primarily of limestone with very little matrix and sporadically cemented, and which cover the Miocene sediments. At the foot of the slope a gully cuts the debris into the underlying Miocene bedrock, causing a sharp break in the slope. We differentiate between two kinds of closed depressions: funnel-shaped and well-shaped dolines. The first kind are numerous and their dimensions can reach a maximum width of 35 m and 20 m in depth, whereas

Photo 3: *Funnel-shaped dolines developed on debris slopes.*

Photo 4: *Surface micromorphology: small rills developed along crack systems.*

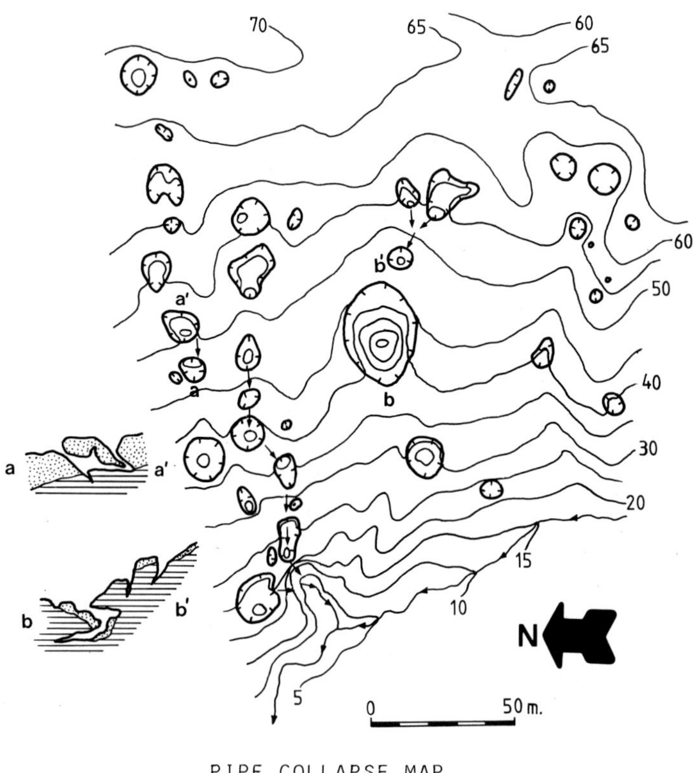

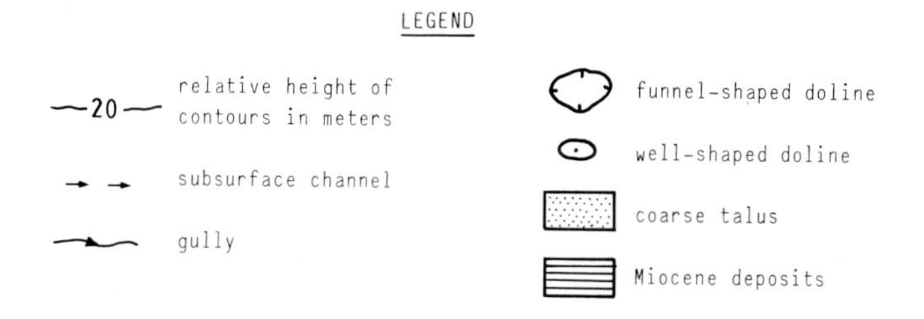

Fig. 4: *Map of pseudodolines developed in coarse talus.*

the second type never go beyond 5 m in diameter, which is less than the depth. Some irregular contour shapes exist as a result of coalescent processes. At the bottom of some funnel-shaped dolines subsidence cones can be observed.

The perimeters of these depressions have in most cases become elongated because of a network of large rills which collect surface water which then infiltrates through the highly permeable debris, reaching the Miocene bedrock where a pipe network is established at the debris-bedrock interface. A few pipes even penetrate the bedrock (fig.4, bb′). The broadening of the subsurface ducts produces a decrease in mechanical resistance in their roofs which gives rise to collapse, a process indicated on the surface by pseudodolines. The pipe network is clearly connected and because of the size of the ducts, a speleological investigation could be carried out in them. The ducts have rectangular sections. The roof and walls of many are covered by white acicular crystals of thenardite (determined by X-ray diffraction). We can observe spacious galleries where the blocks on the floor are in chaotic order. The normal evolution of all these ducts and galleries seems to lead, by continual collapse, to the formation of gully networks.

5 Genetic Considerations

On the alluvial fans gullying and piping processes give rise to the formation of badlands in which round-crested and knife-edged badlands are differentiated, as are turreted ones. The slopes of these ridges are partially covered by saline efflorescences and by an dense crack network (photo 4), the cracks usually being open and with apertures of circa 2 mm, though they can reach up to 8 mm. Inside the polygons there are incipient crack systems, of irregular random orthogonal type (LACHENBRUCH 1962). At times small vertical pipes can be seen at crack intersection points and a network of microrills is generated along the crack openings (HAIGH 1978).

In the lower parts of the main gully that dissects the alluvial fan (fig.2) a 4.8 m thick stratigraphic profile analysis has been carried out, in which the base is Miocene sediments and the alluvial covering is composed of silt and clays with thin intercalations of gravels. We have taken a sample of the Miocene bedrock (A-1) and three of fine sediments (A-2, A-3, A-4) from bottom to top of the profile. In order to characterise the material a granulometric test was carried out on these samples, which show the silt and clay to be the predominant fraction, whereas the sand content fluctuates between 0–24%. X-ray diffraction tests have been performed on the four samples and they all reflect a mineralogy in the following decreasing order: calcite, quartz, illite, chlorite, with traces of feldspars, dolomite, smectite and interstratified clay minerals. This mineralogy reflects the absence of expansive clays, which some authors have argued as a major cause of piping (PARKER 1963, HEEDE 1971, among others).

We have also performed a dispersion-floculation test, and undertaken chemical analyses and measurements of pH, and electrical conductivity on the water extract from saturate paste (RICHARDS 1954). What is clear from all these analyses (tab.1 and 2) is that all the samples are alkalines, with pH very near to 10. The salt content is seen in the electrical conductivity that attains values of up to 25.9 mmhs/cm. There is a high concentration of sodium of between 80–90% of the total cations, giving rise to SAR values of between 48.3 and 66.5, which correspond to 41.1 and 49.1% of exchangeable sodium percentage (ESP). In order to analyse the participation of dispersion processes in the generation and development of piping, dispersion-floculation tests for the fractions 6.3μ and $<2\mu$ have been carried out. The material's dispersibility is reflected in the weight ratios between the disaggregated fractions both with and without dispersants, with values close to 1 being obtained in all

		pH		WATER EXTRACT FROM SATURATE PASTE											
				Water at	CE	meq./liter									
Sample	CO_3Ca %	H_2O	KCl	saturat. %	mmhos/cm 25°	CO_3^0	HCO_3^-	SO_4^4	Cl^-	Ca^{++}	Mg^{++}	Na^+	K^+	SAR	ESP
A-1	28.4	9.2	8.1	57	15.1	-	1.2	88.0	95.6	11.5	6.5	145.0	0.24	48.3	41.1
A-2	46.5	9.9	8.2	31	6.6	-	3.7	55.0	9.5	3.0	1.5	60.0	0.14	54.5	44.1
A-3	41.2	9.6	8.2	44	11.4	-	3.0	58.0	60.8	3.5	2.5	115.0	0.18	66.5	49.1
A-4	31.1	9.2	8.3	52	25.9	-	1.4	158.0	163.8	22.5	15.0	280.9	0.25	65.1	48.6

Tab. 1: *Analysis of water extract from saturate paste.*

SAMPLE	DISPERSANT	FRACTIONS (μ)		WEIGHT RELATION	
		6.3–2	< 2	<2μ	6.3–2μ
A-1	without with	28.97 27.70	40.67 42.70	0.95	1.05
A-2	without with	12.07 11.92	18.02 20.51	0.88	1.01
A-3	without with	39.95 39.31	28.33 29.56	0.96	1.02
A-4	without with	59.80 33.58	12.99 53.50	0.24	1.78

Tab. 2: *Dispersion-floculation test.*

the cases. These chemical-physical characteristics are considered by numerous authors (see PARKER 1963, HEEDE 1971, JONES 1981 and CROUCH et al. 1986) to be a basic factor in the setting off of and development of piping processes. In this area the concentrated ESP values favour material dispersion, since the percolation of water through the cracks produces a decrease in the salts content by leaching although high levels of sodium saturation remain. In this way, the dispersed material moves quite freely and can be displaced along pre-existing hollows, such as cracks or biotic ducts.

6 Conclusions

In spite of some possible similarity of piping mechanism there are different geomorphic implications between the two areas. The piping processes in the alluvial fan area are obviously different from those which give rise to the subsurface network and to the pseudodolines in the slope area, although in both cases the successive collapses that come about as a result of the existence of a subsurface network are going to generate an embryonal gully network. This is, in turn, controlled by the subsurface network. the continual evolution of this collapsing process brings about the disappearance of any traces of a subsurface network. Nevertheless, the piping processes in the alluvial fans can still go on below—the generated gully and the sequence can repeat itself.

In the two cases under study there is another substantial difference in the conditions prior to piping and gullying which has implications for the pattern of gully development. In the alluvial fan area the topography is subhorizontal with no areas of incipient rilling and surface water collection. However, in the slope area there are areas of surface water collection from which runoff filters through the porous slope materials to the bedrock contact, thereby conditioning the direction of the internal ducts and of the future gully.

From the analysis of the two areas studied it can be deduced that, despite their closeness, the generation of pipe networks is due to very different processes, although in both areas their gradual evolution leads to the formation of gully networks.

Acknowledgement

We should like to thank Professor F. López Aguayo and Dr. J. González (Dept. of Cristalography and Mineralogy, University of Zaragoza), for their help in the X-ray diffraction analysis, and Drs. F. Alberto and J. Machín (Exprimental Estation of Aula Dei, C.S.I.C.) for the analyses performed.

References

AGHASSY, J. (1973): Man-induced badlands topography. In: COATES, D.R. (Ed.), Environmental Geomorphology and Landscape Conservation. Vol. **III**. Non-Urban Regions. Benchmark Papers in Geology. Dowden, Hutchinson & Ross, Inc., 124–136.

BRYAN, R. & YAIR, A. (1982) (Eds.): Badlands Geomorphology and Piping. Geobooks. 408 p. Norwich.

CROUCH, R.M., McGARITY, J.W. & STORRIER, R.R. (1986): Tunnel formation processes in the Riverina area of N.S.W., Australia. Earth Surface Processes and Landforms, **11**, 157–168.

DE PLOEY, J. (1974): Mechanical properties of hillslope and their relation to gullying in Central semi-arid Tunisia. Z. Geomorph. Suppl.Bd. **21**, 177–190. Berlin-Stuttgart.

GILMAN, K. & NEWSON, M.D. (1980): Soil pipes and pipeflow a hydrological study in upland Wales. Geobooks, 114 p. Norwich.

HAIGH, M.J. (1978): Microrills and desiccation cracks: some observations. Z. Geomorph., **22**, 457–461.

HARVEY, A. (1982): The role of piping in the development of badlands and gully systems in south-east Spain. In: BRYAN, R. & YAIR, A. (Eds.), Badland Geomorphology and Piping. Geobooks,. 317–335. Norwich.

HEEDE, B.H. (1971): Characteristics and processes of soil piping in gullies. USDA For. Serv. Res. Pap. RM-68, 15 p. Racky Mt. For. and Range Exp. Stn., Fort Collins, Colorado.

JONES, J.A.A. (1981): The nature of soil piping—a review of research. Geobooks, 301 p. Norwich.

JONES, J.A.A. (1982): Experimental studies of pipe hydrology. In: BRYAN, R. & YAIR, A. (Eds.), Badland Geomorphology and Piping. Geobooks, 335-370. Norwich.

LACHENBRUCH, A.H. (1962): Mechanics of thermal contraction cracks and icewedge polygons in permafrost. Geol. Soc. Am., Spec. Paper, **70**, 65 p.

LEOPOLD, L.B., WOLMAN, M.G. & MILLER, J.P. (1964): Fluvial processes in geomorphology. Freeman, 522 p. San Francisco.

PARKER, G.G. (1963): Piping, a geomorphic agent in landform development of the drylands. Internt. Assoc. of Sci. Hydrology, Publ. no. **65**, 103–113.

RICHARDS, L.A. (1954): Diagnosis and improvement of saline and alkali soils. USDA. Agric. Handbook, **60**, 160 p.

SLAYMAKER, O. (1982): The occurrence of piping and gullying in the Penticton glaciolacustrine silts, Okanagan Valley, B.C. In: BRYAN, R. & YAIR, A. (Eds.), Badland geomorphology and Piping. Geobooks, 305–316. Norwich.

Addresses of authors:
M. Gutiérrez, G. Benito
Departamento de Geomorfología y Geotectónica
Facultad de Ciencias
Universidad de Zaragoza
50009 Zaragoza, Spain
J. Rodríguez
Sección Geología de La Rábida
Palos de la Frontera
Huelva, Spain

CATENA SUPPLEMENT 13 p. 61–77 Braunschweig 1988

SEASONAL VARIATION OF EROSIONAL PROCESSES IN THE KAMIKAMIHORI VALLEY OF MT. YAKEDAKE, NORTHERN JAPAN ALPS

H. **Suwa** & S. **Okuda**, Kyoto

Summary

We have investigated the topographic changes of the Kamikamihori Valley of the eastern slope of Yakedake volcano, Northern Japan Alps during a period of over ten years. Erosional processes in the valley depend on the seasonal hydrologic sequence.

The main topographic changes in the valley are caused by rockfall from the side walls and by debris flow along the valley floor. For the quantitative analysis of erosional processes, periodic photographic recording and repeated topographic survey were carried out from 1981 to 1984. From these field observations, it was found that the rockfall from flat walls progresses most actively in early spring when freezing and thawing are repeated on the wall surface, while that from concave walls progresses most actively from mid-spring to summer by convergent surface flow from rapid snowmelt water or heavy rainfall. Talus belts or cones are formed on the foot of flat or concave walls respectively by the accumulation of debris which is supplied from the walls by rockfall. Then a stepped longitudinal profile appears after the rockfall season by the local difference of piling patterns of debris between from flat and concave walls. But the scouring action of debris flows smooths the stepped profile during heavy rainfall, in summer or autumn. The longitudinal profile is alternatively stepped and smooth season by season.

The total volume of debris moved along the valley floor over a whole year by debris flow seems to keep roughly in balance with that supplied from the sidewalls. And a net scour of the valley floor may occur under extraordinary conditions with the occurrence of huge debris flows.

ISSN 0722-0723
ISBN 3-923381-13-1

3-923381-13-1/88/5011851/US$ 2.00 + 0.25

Resumen

Durante un período de más de diez años se han investigado los cambios topográficos producidos en el valle de Kamikamihori, situado en la vertiente oriental del volcán Yakedake, al norte de los Alpes Japoneses. Los procesos de erosión que tienen lugar en el valle dependen de la secuencia hidrológica estacional.

Los principales cambios topográficos del valle están causados por la caída de rocas de las paredes laterales y por los flujos de derrubios a lo largo del fondo del valle. Desde 1981 a 1984 se llevó a cabo un análisis cuantitativo de los procesos de erosión mediante la toma periódica de fotografías y la repetición de controles topográficos. Estas observaciones de campo mostraron que la caída de rocas desde las paredes lisas del valle es más activa al principio de la primavera, cuando los procesos de hielo y deshielo se repiten frecuentemente en la superfície de la pared, mientras que en las paredes cóncavas la caída de rocas tiene su período más activo desde la mitad de la primavera hasta el verano, a causa de la escorrentía superficial procedente de la fusión de la nievo o de los chubascos. La acumulación de los derrubios producidos por la caída de rocas resulta en una alineación de taludes al pie de paredes lisas, mientras que en las paredes cóncavas se forman series de conos. Asi, después de la estación en que se produce la caída de rocas, aparece un perfil longitudinal escalonado por la diferencia local del tipo de acumulación de derrubios entre las paredes lisas y las cóncavas. Pero la erosión producida por los flujos de derrubios durante los chubascos de verano y de otoño alisa el perfil escalonado. Por tanto el perfil longitudinal aparece alternativamente escalonado y liso de una estación a otra.

El volumen total de derrubios movilizados a lo largo del fondo del valle durante un año a causa de los flujos de derrubios parece mantener un balance aproximado con el material suministrado por las paredes laterales. Una erosión neta del fondo del valle puede tener lugar en condiciones extraordinarias, como cuando se producen grandes flujos de derrubios.

1 Introduction

In most mountainous regions of Japan, there is a distinct seasonal cycle in the meteorological and hydrological conditions. The accumulation of debris on valley bottoms and the flushing out of the debris mass from the valleys proceed usually according to some seasonal sequence. In some mountains with steep slopes and strongly weathered rocks or unconsolidated deposits, the rate of accumulation and debris flush out is so large that we can investigate the seasonal variations of the erosional and depositional processes by field observation in only a few years (SUWA et al. 1980, 1981, 1983, 1985a, OKUDA et al. 1980, 1984).

We studied the mass balance and transport action of debris in the Kamikamihori Valley on the eastern slope of Mt. Yakedake in the Northern Japan Alps. In 1981 and 1982, we observed the supply rate of debris fragments while pursuing the normal field observation work on the debris flow phenomena. We repeated valley profile surveys to study the relationship between debris mass balance and changes in the valley bottom profiles.

2 Measurement of Rockfall from Valley Walls

We started the measurement of rock fall from the sidewalls of the Kamikamihori Valley in the study area, shown in fig.1, where many instruments were set to observe hydrological elements and motion of debris flow every summer. The explanation of the observation system for debris flow is not included in this paper, because it has already been published (OKUDA et al. 1980, SUWA et

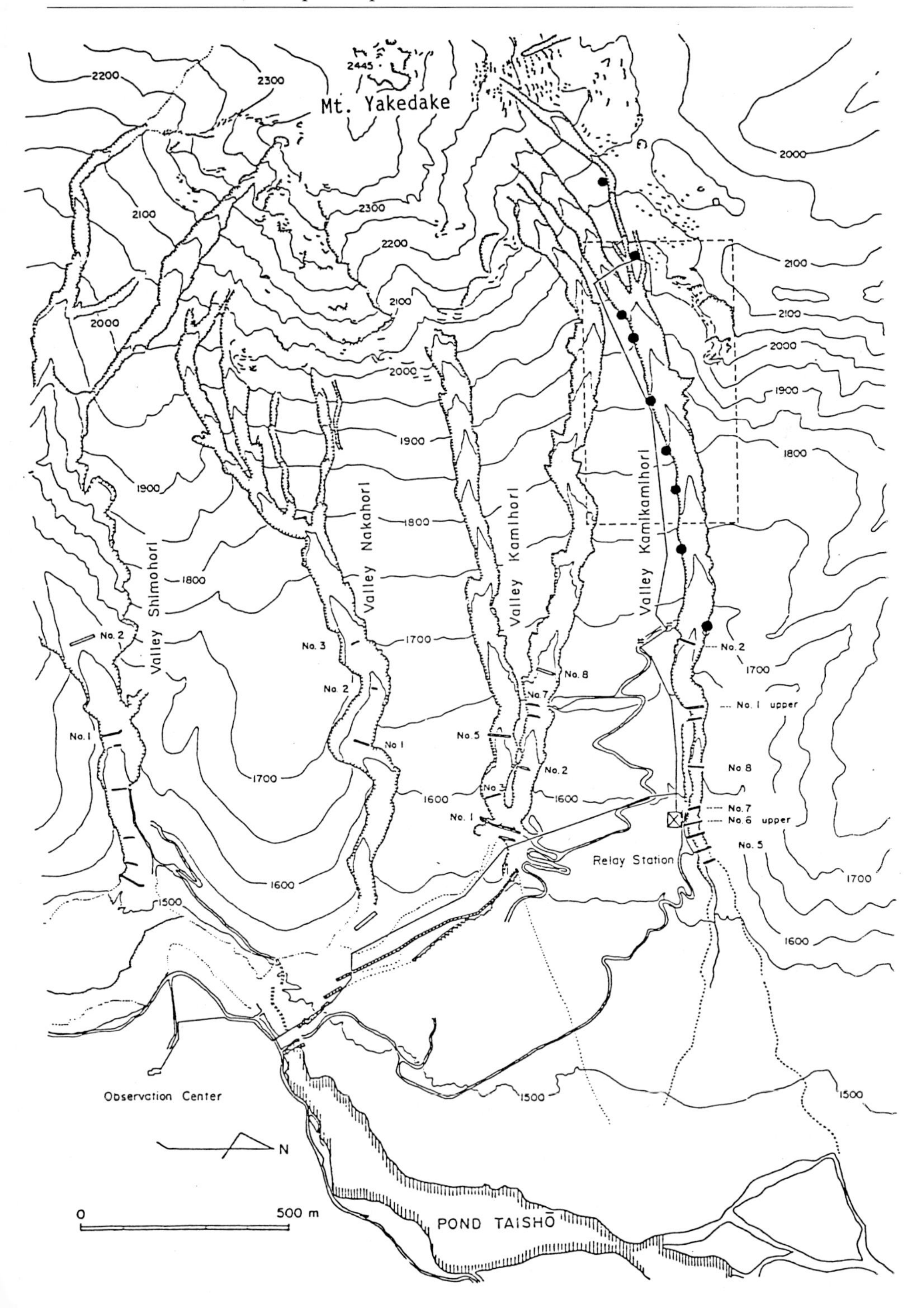

Fig. 1: *Location of the study area.*

The broken line rectangle shows the main study area, where the sidewall erosion is very active. Solid circles show the rock fall survey slopes.

Photo 1: *Two shots of sidewall No.4.*

(a): June 17, 1981, corresponding to the initial date in fig.4. (b): Sept. 29, 1982, corresponding to the final date in fig.4.

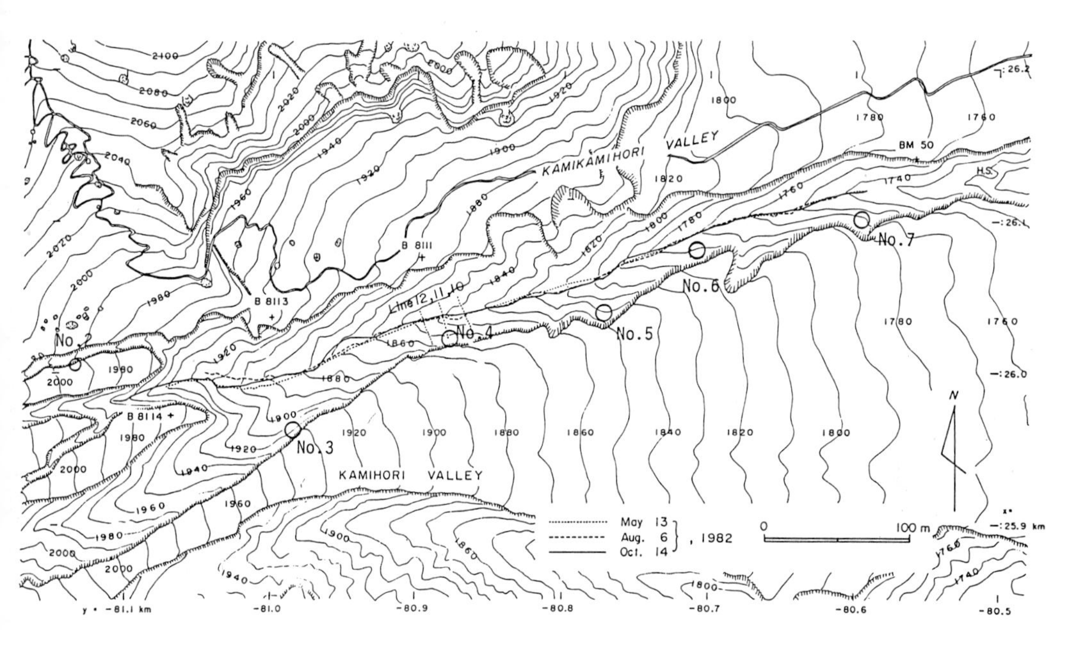

Fig. 2: *Detailed map of the study area in fig.1.*

Survey lines are drawn for a longitudinal profile along the talweg and three lateral cross sections (line 10, 11, 12) of the valley bottom. Cross symbols show the bench mark points for the topographical survey.

al. 1985b).

Rock fall was measured at locations of solid circles in fig.1. A detailed map of the study area in the rectangle is shown in fig.2, and three survey lines, 10, 11 and 12, were set to study the changes in the cross sections of the valley. The bench mark points for topographic surveys are shown by cross symbols in the figure.

Photographs were taken over the period July 1981 to Sept. 1982 from fixed points (the solid cricles in fig.1), to record the change in the sidewall surface from rock fall. An observer took photos of the same sidewall surface with the same field of view about every month. But at the point B8111 (see fig.2 and fig.3), a motor driven 35 mm camera was set up to take the photos of the slope surface of the opposite sidewall every two hours.

A set of the photos of sidewall No.4 (photo 1) shows many square targets (20 cm × 20 cm) hanging on ropes which were used as a scale in the analysis. By comparing the sequential photos, we can detect changes in the sidewall surface caused by the rock fall during the period between the dates of two photos. Fig.4 shows the features of the rock fall over the whole observation period from June 17, 1981 to Sept. 29, 1982.

The two photos were compared by overlapping two transparent films of two enlarged photos. Only the traces of vanishing and remaining boulders larger than 20 cm in diameter are illustrated in fig.4 because the resolving power of the original photo is about 20 cm at the sidewall surface. The changes in other sidewalls were also investigated similarly

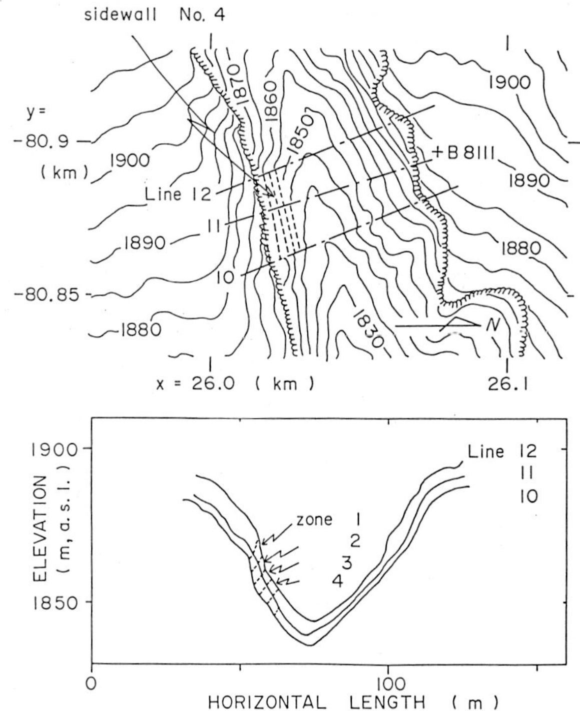

Fig. 3: *Top: plan of the sidewall No.4 and survey lines for cross sectional profile of valley bottom. bottom: profiles of the valley derived from topographical map.*

to those on sidewall No.4, but only the traces of boulders larger than 45 cm in diameter were pictured owing to the limited accuracy of the photos. The total volume of fallen boulders was estimated from the numbers and size of the boulders.

3 Seasonal Variations in Rock Fall

Rock fall is controlled by hydrological conditions at the sidewall surface and also by the slope form and materials.

3.1 Rock Fall from Flat Walls

The sequence of rock fall can be recorded by the cumulative volume of the fallen boulders. An example of rock fall from the flat sidewall No.4 is shown in fig.4. The wall was devided into four zones as shown in the figure, and the numerical records of fallen boulders and rainfall conditions are given in tab.1. The data for rainfall at Taishoike pond near the valley were provided by the Azusagawa office of Tokyo Denryoku Co. Ltd. The area ratio of fallen boulders is defined as the ratio of the total exposed area of the traces of the vanishing boulders to the whole area of the wall in the field of view of the photos. The rock fall is recorded in fig.5 as the cumulative number, projected area and volume of fallen boulders relative to the total value through the whole observation period.

The cumulative percent of three values in fig.5 tells us that rate of rock fall is greatest from April to May when the temperature rises above 0°C and freezing and thawing are repeated most fre-

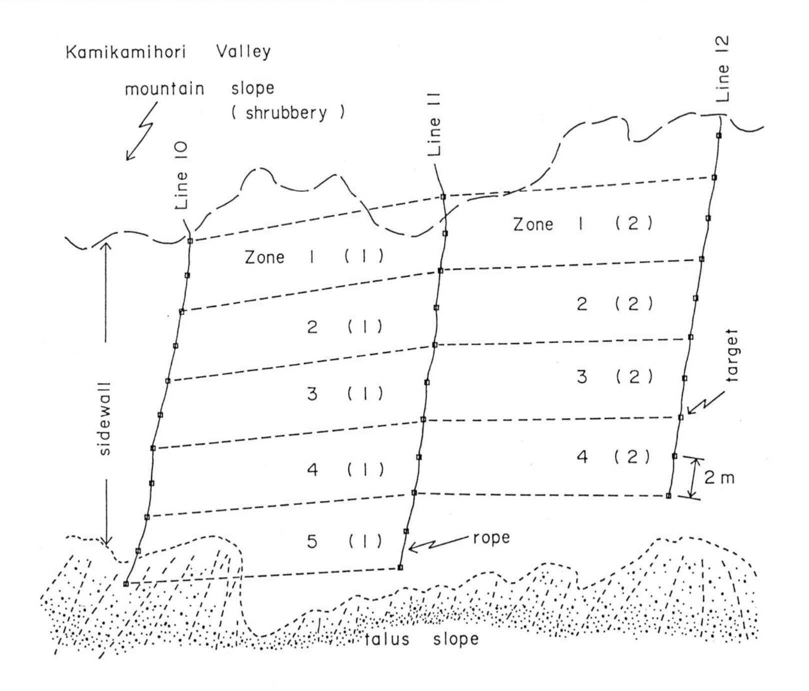

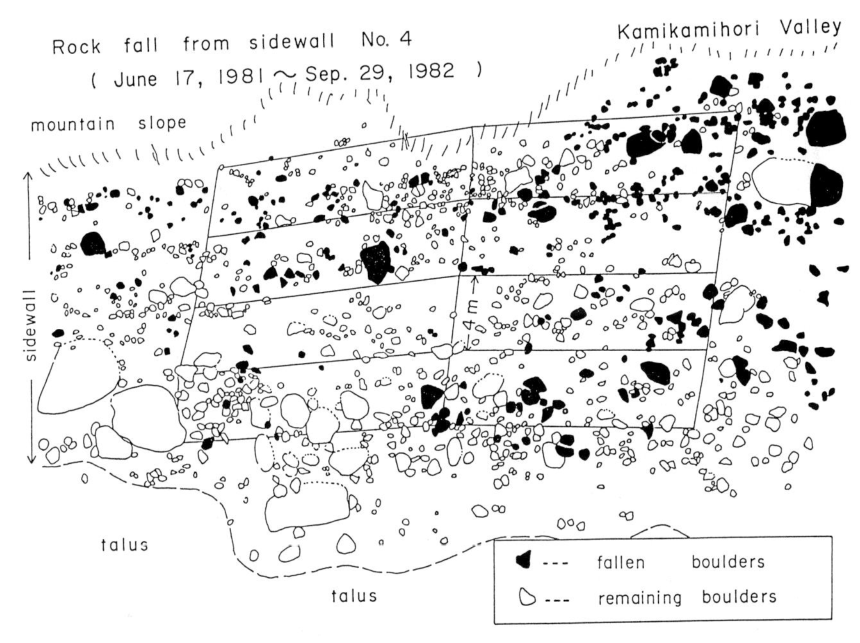

Fig. 4: *Top: Hanging ropes for scale standard and zoning on sidewall No.4 with square targets (20cm × 20cm). The distance between lines 10, 11 and 12 is 15 m. Bottom: Rock fall from sidewall No.4 (flat slope), which occurred during the observation period from June 17, 1981 to Sept. 29, 1982.*

date	fallen boulder				rainfall condition (mm)				
	no.	area ratio (%)	volume (m^3)	size of largest boulder (m)	total rainfall*	maximum daily rainfall*	mean of daily rainfall*	maximum hourly rainfall**	maximum 10 min. rainfall**
1981 June 17									
	23	1.64	3.83	1.2	880.5	133.5	26.7	20.5	15.5
July 20									
	9	0.72	1.68	1.0	81.5	19.5	3.0	9.5	4.5
Aug. 14									
	1	0.04	0.06	0.5	498.5	110.0	16.1	26.0	11.0
Sept. 14									
	5	0.17	0.28	0.7	331.5	84.0	11.1	12.0	4.5
Oct. 14									
	5	0.29	1.34	0.8	(655.0)	(48.0)	(3.3)	-	-
1982 Apr. 27									
	81	6.23	18.99	2.4	101.0	30.0	5.6	-	-
May 16									
	2	0.42	1.47	1.3	224.5	73.0	7.0	11.5	3.5
June 17									
	0	0	0	0	174.5	34.0	5.1	10.0	6.0
July 21									
	14	0.69	1.51	1.2	204.0	46.5	13.6	13.5	5.5
Aug. 5									
	4	0.25	0.49	0.8	189.0	58.0	6.8	15.0	6.0
Sept. 2									
	2	0.36	1.15	1.1	352.5	86.0	13.1	18.5	5.5
Sept. 29									

* at Taishoike-pond (by Tokyo Denryoku Co. Ltd)
** at R_4 (source area of Kamikamihori valley)
() precipitation, total precipitation, 655 mm, of which 60 mm fell as rain, the remainder as snow

Tab. 1: *Sequence of rock fall from sidewall no. 4 and rainfall conditions.*

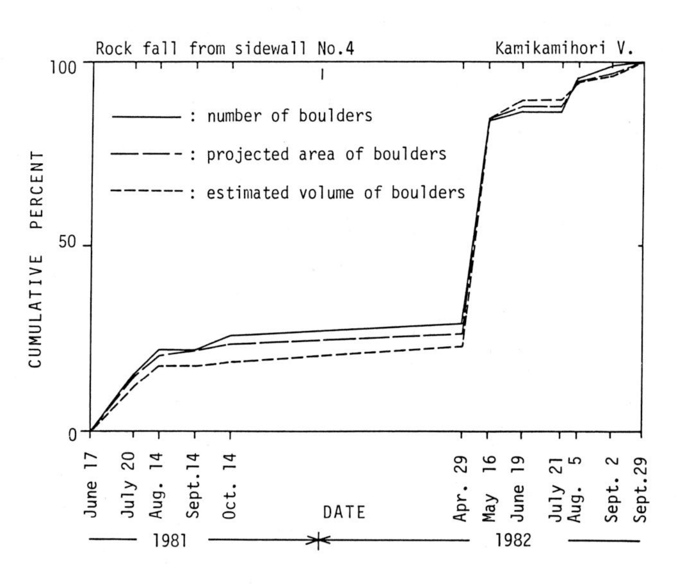

Fig. 5: *Sequence of rock fall shown as the cumulative number, projected area and volume of fallen boulders from sidewall No.4.*

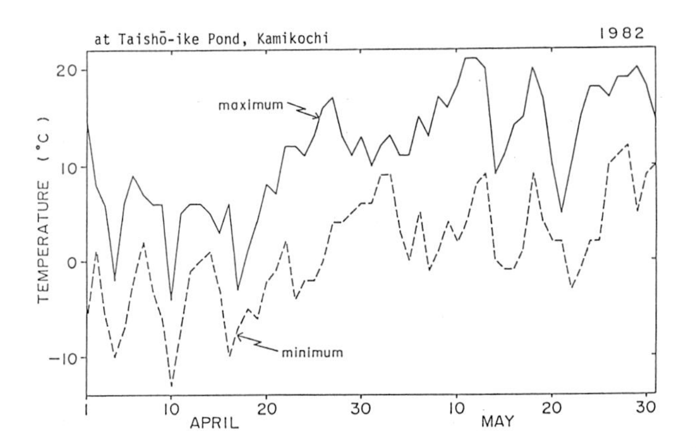

Fig. 6: *Maximum and minimum temperature during April and May 1982 at Taishoike pond, observed by Tokyo-Denryoku Co. Ltd.*

quently, as shown by the air temperature record in fig.6. It should be noted that the air temperature in the study area would be approximately 2°C lower than the one at Taishoike pond in fig.6 because of an altitude difference of about 400 m between the two places. The second largest rate of rock fall appeared from June 17th to July 20th, 1981, when heavy rainfall occurred (tab.1). The third largest rate of rock fall occurred from July 20th to Aug. 14th, 1981, when rainfall is small but the sidewall surface becomes dry and active wind erosion of the matrix proceeded thereby loosening the boulders because of strong wind along the valley.

Besides the rapid progress of rock fall mentioned above, heavy rainfall and strong winds from the typhoon on Aug. 1st 1982, basal erosion of the sidewall and strong vibrations from a debris flow on Sept. 20th 1982 stimulated rock fall from the wall.

The four zones of the sidewall (fig.4) show distinctive patterns in the rock fall sequence. Rock fall proceeded actively in zones 2 or 3 influenced by vibration of trees from strong winds, while some active rock falls were brought about in zone 4 by basal erosion of the wall or strong vibrations caused by a passing debris flow.

3.2 Rock Fall from a Convergent Sidewall

Some sidewalls are concave upward, collecting rainfall or snowmelt. Surface erosion is apt to occur at such sidewalls by surface runoff. We call such walls "convergent sidewalls".

The sequence of rock fall in convergent wall No.7 was recorded by the interval photo method. Photo 2 shows one shot of this record. From such photos the changes in the wall surface are depicted as shown in fig.7. At the convergent sidewall, the traces of fallen boulders are concentrated in a special band zone as found in this figure.

The seasonal sequence of rock fall from the convergent wall No.7 is shown as the change in the cumulative percent of the trace area of fallen boulders in fig.8 compared to the results from flat sidewall No.4.

Photo 2: *Sidewall No.7 (convergent slope) on Sept. 28, 1982.*

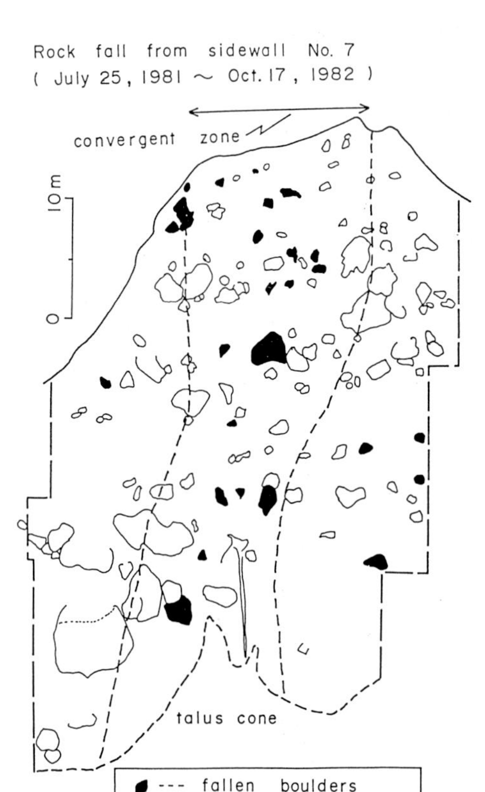

Fig. 7: *Rock fall from sidewall No.7 (convergent slope) as found in photo 2 during the observation period from July 25, 1981 to Oct. 17, 1982.*

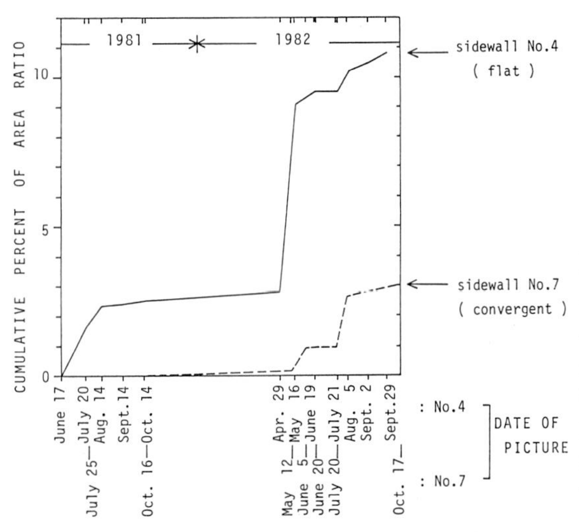

Fig. 8: *Comparison of rock fall sequences from sidewall No.4 (flat slope) and sidewall No.7 (convergent slope) by their cumulative area ratios of fallen boulders (ratio of the projected area of fallen boulders to the whole survey area of the sidewall).*

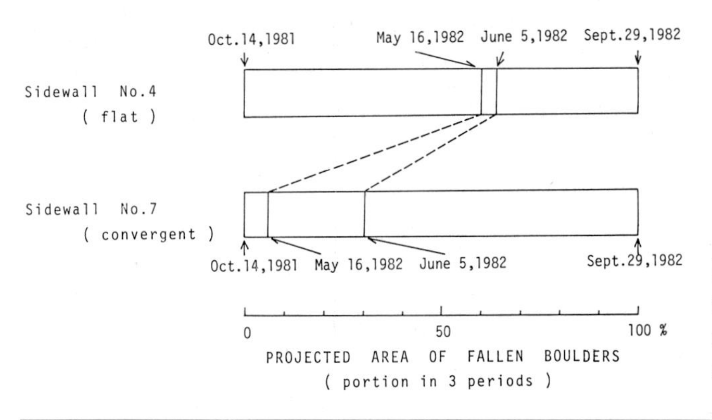

Fig. 9: *Comparison between three periods of rock fall for two different sidewalls.*

zone	volume of fallen boulders $V_f (m^3)$	ratio (projected area of total boulders / area of sidewall) (P)	total volume of fallen debris $V = V_f / P (m^3)$	mean retreat of sidewall $V/A (m)$
1	7.0	0.184	38	0.32
2	11.5	0.179	64	0.57
3	3.7	0.150	25	0.21
4	5.0	0.280	18	0.15
total	27.2	-	145	-
The survey slope of sidewall No. 4; June 17, 1981 June 17, 1982				

Tab. 2: *Volume of fallen boulders and estimated mean retreat of sidewall for each zone in fig.4.*

From this figure, it seems that the largest rate of rock fall from the convergent wall appeared during the heavy rainfall of the typhoon on Aug. 1st, 1982; the second largest during the snowmelting season from May to June.

Projected area of fallen boulders is shown in fig.9 to compare different progress periods of rock fall from different sidewalls. In this figure, projected area of fallen boulders relative to the total area of traces through the whole observation period is shown by the partition of rock fall occurrence in three specified periods.

In the first period, from the middle of Oct. 1981 to the middle of May, 1982, most rock fall occurred in late spring because of repeated freezing and thawing. In the second period, from the middle of May to the beginning of June, 1982, rock fall was caused mainly by wind erosion in the dry flat sidewalls, while by surface runoff in the wet and convergent sidewalls. In the third period, from the beginning of June to the end of September, 1982, active rock fall was caused by heavy rainfall, strong typhoon wind or by basal erosion and vibration from a debris flow. From this figure, it is noticed that rock fall occurred most actively in the first priod for the flat sidewalls, while in the third period for the convergent sidewalls.

4 Side Erosion and Retreat Rate of Sidewall

We can estimate the total volume of fallen debris (V) from the survey slope of sidewall using the rock fall data as $V = V_f/R$, where (V_f) is the volume of fallen boulders and (R) is the volume ratio of the boulders to the total debris. In practice, we can hardly know the real value of R only from the photo data. From the photo film, we can measure the ratio (P) of the projected area of the boulders to the area of survey slope.

From a simple consideration on a model slope with homogeneous and isotropic material composition, it is shown theoretically that the expected value of P is slightly smaller than R and we use the measurable value (P) for the approximate value of (R).

Then, we can estimate the total volume of fallen debris from the equation $V = V_f/P$.

The results are arranged in tab.2 for each zone in sidewall No.4. The mean retreat rate of the sidewall can be also calculated as the total eroded volume (V) divided by the total wall area (A); the values of the retreat rate averaged over the whole observation period amounted to the order of 10^{-1} m/year.

From the above value for the eroded volume 145 m^3 from the sidewall N.4 of 30 m length, we can estimate the annual volume of the debris supplied from the whole sidewall in the study area in fig.1 (500 m length $\times$ 2 sidewalls). The final value $5 \times 10^3 m^3$ was obtained from the above method. This value is of the same order as the one for debris volume which is removed from the valley by debris flow flush in one year (SUWA et al. 1985b), excluding an extraordinary year when a very large debris flow carried away a great amount of debris in the order $10^4 m^3$. This was estimated by independent field surveys. As for the mass balance of the debris in the valley, debris fragments are supplied fro sidewalls through rock fall mainly in late spring and early summer. The debris accumulated on the valley bed is carried away mainly in summer and autumn through the flushing action of debris flows.

5 Seasonal Changes in Valley Profiles

Sequential accumulation and flushing of debris in the valley bring about a seasonal change in the cross sectional form of the valley. We have surveyed the profile of the valley repeatedly to monitor these seasonal changes.

In winter when the surface of the sidewall is frozen and the valley bed is covered by a thick snow layer, the movement of debris is almost stopped and no change in the cross section occurs in the valley.

In late spring and early summer, rock fall proceeds rapidly and a large amount of debris from the sidewalls changes the valley profile remarkably. Supplies of debris from sidewalls produce talus along the foot of the walls, but the talus development takes different forms between the flat and convergent walls.

Along the foot of a flat sidewall, a talus belt with a rather uniform cross section develops, while at the foot of the convergent sideall, a typical talus cone develops with the main axis transverse to the valley.

The contrast between the two types of talus is shown in fig.10 with a map of the valley (top) and a sketch of the two cross sections (bottom).

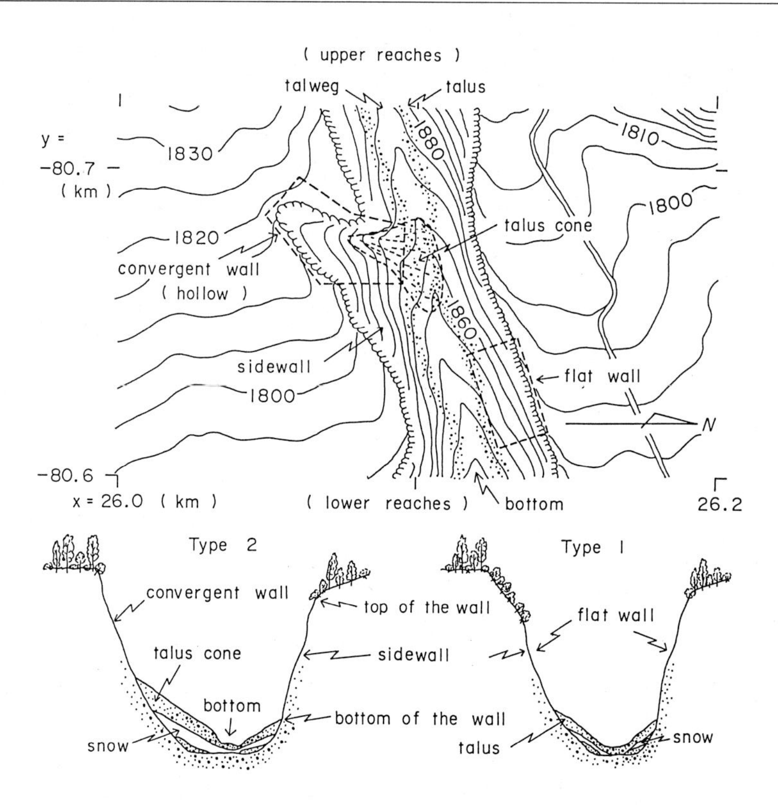

Fig. 10: *Classification of sidewall and the depositional structure over the valley bottom: plan (top) and cross sections (bottom).*

The talus cones under the convergent sidewall produces steps in the longitudinal profile along the talweg of the valley bottom after the debris accumulation period (fig.11). However, during the heavy rainfall season in summer and early autumn, debris flows change the valley profile remarkably by removing the accumulated debris. A small debris flow on Sept. 20, 1982 smoothed the profile of the upper region, keeping the stepped profile partly by weak scouring, as shown in fig.11a. Larger debris flows on Aug. 17 and on Sept. 4, 1978 smoothed the valley profile almost completely by strong scouring action over the whole reach, as

Fig. 11: *Changes in the longitudinal profile of the valley bed in the upper reach of survey area, and the locations of the center line of convergent wall (Left 1∼L. 8 and Right 1∼R.4) and those of talus cone (1∼5).*

Convergent wall with dot and broken line froms a remarkable talus cone. A small-scale debris flow occurred on Sept. 20, 1982 and a large-scale debris flow occurred on Aug. 17, 1978. The effects of the former and the latter events appear in fig.11a and in fig.11b respectively. →

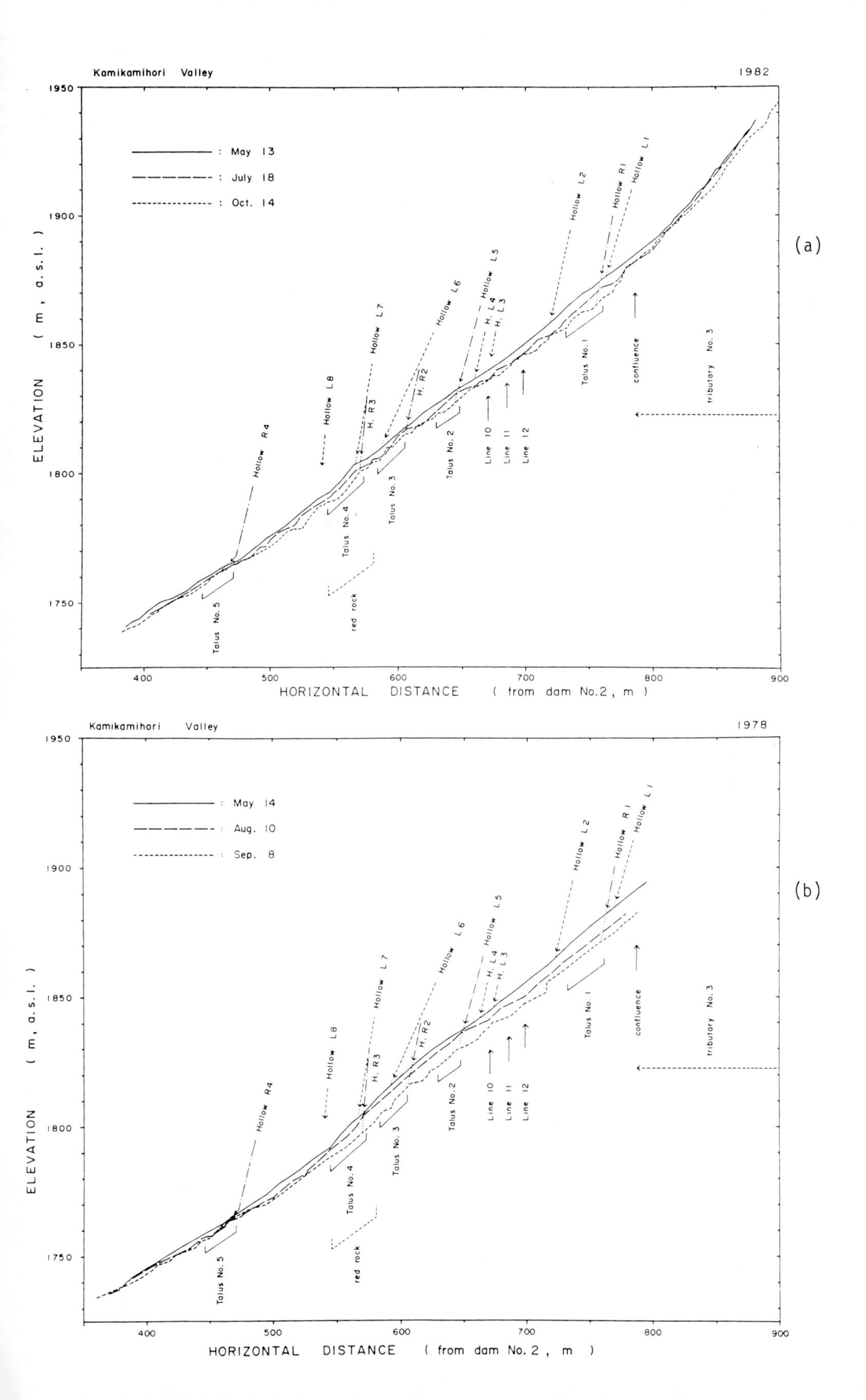

Kamikamihori Valley
1982
: May 13
: July 18
: Oct. 14
ELEVATION (m, a.s.l.)
HORIZONTAL DISTANCE (from dam No.2, m)
Hollow R4
Hollow L8
Hollow L7
H. R3
Hollow L6
H. R2
Hollow L5
H. L4
H. L3
Hollow L2
Hollow R1
Hollow L1
Talus No.5
Talus No.4
Talus No.3
Talus No.2
Talus No.1
red rock
Line 10
Line 11
Line 12
confluence
tributary No. 3
(a)
Kamikamihori Valley
1978
: May 14
: Aug. 10
: Sep. 8
ELEVATION (m, a.s.l.)
HORIZONTAL DISTANCE (from dam No.2, m)
(b)

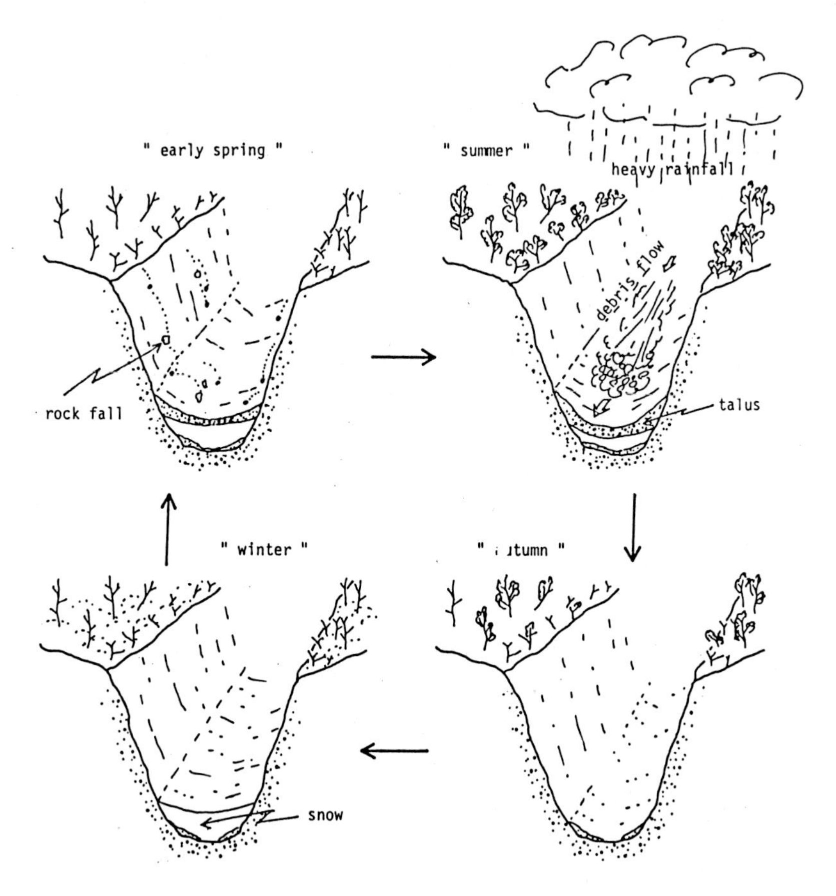

Fig. 12: *Cyclic and seasonal changes in the depositional and erosional conditions at the upper reaches of a valley.*

shown in fig.11b.

6 Conclusion

In conclusion, in some valleys of high mountains in Central Japan, active rock fall, induced firstly by freezing and thawing in early spring and secondly by surface runoff with heavy rainfall in early summer, supplies a large amount of debris from sidewalls to valley bottoms. In summer and autumn, debris flows caused by heavy rainfall carry away the accumulated debris on the valley bottoms in a short time by a very strong transport capacity in the longitudinal direction. Owing to the difference of talus forms between flat and convergent sidewalls, a stepped profile in the longitudinal direction appears after debris accumulation, but flushing action of large debris smoothes this stepped profile. Through the above erosional and depositional processes, a distinctive seasonal change in the valley profile would proceed as shown schematically in fig.12 according to the annual hydrological sequence.

Acknowledgement

The authors thank Mr. T. Shikata of the Department of Forest Conservation of the Kyoto Prefectural Government and Dr. S. Kitazawa of Shinshu University for their cooperation and useful discussions, and the Matsumoto Erosion Control Works of the Ministry of Construction for its cooperation in the field observation of debris flows at Mt. Yakedake.

References

OKUDA, S., SUWA, H., OKUNISHI, K., YOKOYAMA, K. & NAKANO, M. (1980): Observation on the motion of a debris flow and its geomorphological effects. Z. Geomorph. N.F., Suppl. **35**, 142–163.

OKUDA, S. & SUWA, H. (1984): Some Relationships between Debris Flow Motion and Micro-Topography for the Kamikamihori Fan, North Japan Alps. In: BURT, T.P. & WALLING, D.E. (Eds.), Catchment Experiments in Fluvial Geomorphology. Geo Books Norwich, 447–464.

SUWA, H., OKUDA, S. & YOKOYAMA, K. (1973): Observation System on Rocky Mudflow. Bull. Disast. Prev. Res. Inst., Kyoto Univ., **23**, 59–73.

SUWA, H. & OKUDA, S. (1980): Dissection of valleys by debris flows. Z. Geomorph. N.F. Suppl. **35**, 164–182.

SUWA, H. & OKUDA, S. (1981): Topographical Change Caused by Debris Flow in Kamikamihori Valley, Northern Japan Alps. Trans. Jap. Geomorph. Union, **2(2)**, 343–352.

SUWA, H., SHIKATA, T. & OKUDA, S. (1983): Topographical Change on the Sidewall and in the Valley Bottom of the Kamikamihori Valley of Mt. Yakadake. Annuals of Disast. Prev. Res. Inst., Kyoto Univ., 26B-1, 413–433 (in Japanese).

SUWA, H., SHIKATA, T. & OKUDA, S. (1985a): Topographical Changes in the Kamikamihori Valley on the Yakedake Volcano, Japan. Shin-Sabo, **37(5)**, 136, 14–23 (in Japanese).

SUWA, H. & OKUDA, S. (1985b): Measurement of Debris Flows in Japan. In: Proc. IVth Inter. Conf. and Field Workshop on Landslides, Tokyo. 391–400, 1985.

Addresses of authors:
H. Suwa
Disaster Prevention Research Institute
Kyoto University
Uji City
Kyoto Prefecture, 611
Japan
S. Okuda
Faculty of Science
Okayama University of Science
Ridai-cho 1-1
Okayama City, 800
Japan

Peter D. Jungerius (Ed.):

Soils and Geomorphology

CATENA SUPPLEMENT 6 (1985)

Price DM 120,-

ISSN 0722-0723 / ISBN 3-923381-05-0

It was 12 years ago that CATENA's first issue was published with its ambitious subtitle "Interdisciplinary Journal of Geomorphology – Hydrology – Pedology". Out of the nearly one hundred papers that have been published in the regular issues since then, one-third have been concerned with subjects of a combined geomorphological and pedological nature. Last year it was decided to devote SUPPLEMENT 6 to the integration of these two disciplines. Apart from assembling a number of papers which are representative of the integrated approach, I have taken the opportunity to evaluate the character of the integration in an introductory paper. I have not attempted to cover the whole bibliography on the subject: an on-line consultation of the Georef files carried out on 29th October, 1984, produced 3627 titles under the combined keywords 'geomorphology' and 'soils'. Rather, I have made use of the ample material published in CATENA to emphasize certain points.

In spite of the fact that land forms as well as soils are largely formed by the same environmental factors, geomorphology and pedology have different roots and have developed along different lines. Papers which truly emanate the two lines of thinking are therefore relatively rare. This is regrettable because grafting the methodology of the one discipline onto research topics of the other often adds a new dimension to the framework in which the research is carried out. It is the aim of this SUPPLEMENT to stimulate the cross-fertilization of the two disciplines.

The papers are grouped into 5 categories: 1) the response of soil to erosion processes, 2) soils and slope development, 3) soils and land forms, 4) the age of soils and land forms, and 5) weathering (including karst).

P.D. Jungerius

CATENA SUPPLEMENT 13 p. 79–95 Braunschweig 1988

SOME ASPECTS OF THE GEOMORPHIC PROCESSES TRIGGERED BY AN EXTREME RAINFALL EVENT: THE NOVEMBER 1982 FLOOD IN THE EASTERN PYRENEES

F. **Gallart** & N. **Clotet-Perarnau**, Barcelona

Summary

During the 6th and 7th November, 1982, an unusual rainfall event occurred in the Eastern and Central Pyrenees. The rain lasted for almost 48 hours, and the total amount of precipitation reached 556 mm in La Molina (near Puigcerdà).

The authors studied the consequences of the flood in the high basin of the LLobregat and Cardener rivers. The most important changes were analysed briefly after the event, and a detailed inventory of the hillslope phenomenon was made a few months later.

In spite of the large rainfall total, the precipitation intensity scarcely reached 10 mm per hour; this could be the cause of the general absence of new rills or gullies on the hillslopes; while the long duration of the effective rain resulted in very high water stages in the main rivers. The most important processes were scouring of most of the stream beds, especially in the low order channels, and a relatively high number of mass movements on the hillslopes. A few large mass movements occurred in some areas where there was evidence of earlier mass movement.

The inventory of mass movements showed three main results: most of the shallow slips occurred in old man-made agricultural terraces; slump-type slides seem to be promoted by forest cover; and the number of the mass movements increases with the 7th. power of total rainfall.

Resumen

Durante los dias 6 y 7 de noviembre de 1982 cayeron en el Pirineo oriental y central unas lluvias fuera de lo usual. La precipitación duró casi 48 horas y en el observatorio de La Molina (cerca de Puigcerdá) se recogieron un total de 556 mm. Se han estudiado las consecuencias de la crecida en las cabeceras de las cuencas de los ríos LLobregat y Cardener. Los cambios más importantes fueron analizados con rapidez poco después del acontecimiento, y unos meses más tarde se llevó a cabo un inventario de los fenómenos ocurridos en las vertientes.

A pesar del gran volumen de agua caído la intensidad de la precipitación

ISSN 0722-0723
ISBN 3-923381-13-1

3-923381-13-1/88/5011851/US$ 2.00 + 0.25

a penas llegó a los 10 mm/h. Este hecho podría explicar la ausencia de nuevas cárcavas o abarrancamientos en las vertientes, en contraste con lo ocurrido en los lechos fluviales, los cuales estuvieron sometidos a los efectos de una notable crecida a causa de la persistencia de la lluvia. Los procesos más importantes fueron la erosión de la mayoría de lechos fluviales, especialmente en los canales de orden inferior, y un número relativamente alto de movimientos en masa en las vertientes. Unos pocos movimientos en masa de gran magnitud tuvieron lugar en algunas áreas con evidenca de antiguos movimientos. El inventario de los movimientos en masa presenta tres resultados principales: la mayoria de deslizamientos poco profundos ocurrieron en antiguas terrazas de cultivo; los deslizamientos rotacionales aparecen más frecuentemente bajo el bosque; el número de movimientos en masa alrededor de las estaciones meteorológicas crece según la séptima potencia del total de la precipitación.

Introduction

The aim of the present paper is to analyse the geomorphic consequences of a continuous rainfall event, in relation to geographical variables (altitude, topographic gradient, lithology, land use) and spatial distribution of the rain, and to contrast the processes with those associated with ordinary meteorological events.

The flood occurred during a study of the sediment dynamics of the Llobregat basin (CLOTET & GALLART 1986). A few weeks after the extreme event, a survey of the main phenomena was made (CLOTET & GALLART 1983) by inspection of the most important mass movements, and by detailed mapping of some reaches in the main channels. Some months later, a field inventory of the hillslope processes in an area of 1250 km^2 was carried out (CLOTET & GALLART, in press). At the present time, the aim of our research program is to analyse the importance of such infrequent events in the origin of badland areas.

1 Characteristics of the Studied Area

The upper Llobregat and Cardener basins lie in the southern Pyrenees (fig.1), and are composed of sedimentary rocks of Devonian to Oligocene age. The main peaks, with heights of about 2500 m, are on limestones, while nonmarine mudstones and marine marls, with gentler slopes, outcrop in valleys and depressions. The annual precipitation varies between 800 and 1500 mm and shows a very irregular distribution during the year; the mean annual temperature is about 10°, with minima of −12° and maxima of 30°.

Climate, the extent of calcareous rocks, and human activity are the main controls on vegetation cover. Below 1600 m the forests of *Pinus sylvestris* very often replace the *Buxo-Quercetum pubescentis* association, which is only present in some remnants. Above 1600 m the *Pulsatillo-Pinetum uncinatae* association is dominant, with a grass layer and some groups of *Juniperus communis*. The forest covers only a 52% of the area since below 1300 m most of the gentler slopes are terraced for agricultural use, although nowadays many of these are abandoned. The rocky slopes are often covered by a scrub vegetation of bushes of *Buxus sempervirens*, and most

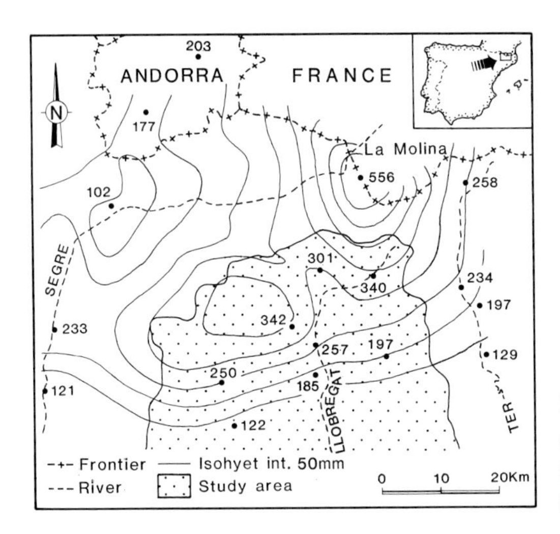

Fig. 1: *Location of study area. Figures and isohyets show the total amount of rainfall for the period 6th–8th November 1982, inclusive.*

of the high summits are rocky or covered by grasslands. The population density is low on account of the mountainous character of the area, with only coal mining and indutrial activities counteracting the trend towards depopulation.

2 The Rain Event

During the 6, 7 and 8 November 1982, large amounts of rain were collected in most of raingauges of the central and eastern Pyrenees (fig.1). The highest amounts were 556 mm in La Molina (near Puigcerda) and 650 mm in Refugio de Moriz (near Huesca). The 24-hour rainfall totals for 7th November exceeded the highest records for most of the stations, but the short lengths of record (less than 30 years) make it difficult to estimate the return period for the storm. After MARTIN-VIDE & RASO (1983), the return period is 55 years for the 24 hr rainfall collected in Berga (135 mm, 14-year record length), and 100 years for La Pobla de Lillet (168 mm, 24-year record length), both within the study area.

Unfortunately, there is no record of rainfall intensity, only the data from La Molina, with a reading every 6 hours, can give some detail of the event. This station demonstrates that the rain started at 06.00 hours on 6th November with increasing intensities between 5 and 10 mm per hr, at 24 hours, the intensity increased to 17 mm per hr and continued with values between 13 and 21 mm per hr until 24.00 hours on the following day, 7th November. On the last day, the rain lasted for only 7 hours, with a mean intensity of 12 mm per hr. The duration of the rain was nearly the same for all the sites, and therefore approximate mean intensities can be estimated from those at La Molina.

3 Runoff Processes, Flood Discharge and Channel Change

According to several lines of evidence, the flood originated on the hillslopes as a result of saturation overland flow of long duration but low erosive power.

The distribution of flood peaks es-

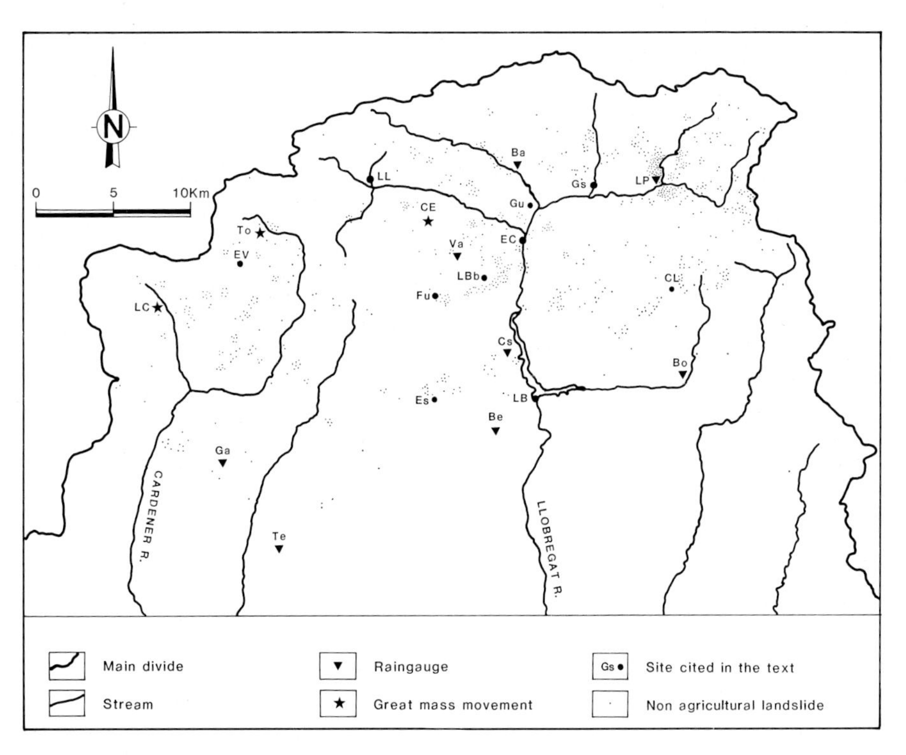

Fig. 2: *Location of the sites cited in the text, and of natural mass movements (failures of artificial slopes on farming terraces excluded).*

Ba = Bagà, Be = Berga, Bo = Borredà, CE = Clot d'Esquers, CL = Can Lloveres, Cs = Cercs, EC = El Collet, Es = Espinalbet, EV = El Verd, Fu = Fumanya, Ga = Guilanyà, Gs = Gavarrós, Gu = Guardiola, LB = La Baells, LBb = La Barrumba, LC = La Coma, LL = Llúria, LP = La Pobla de Lillet, Te = Tentellatge, To = Torrentsenta, Va = Vallcebre.

timated by the Manning equation in 18 tributaries of the Gavarrós basin (22 km^2) (location on fig.2) is closely related to catchment size; the regression equation takes the form

$$Q = 5.47 \times A^{0.828}$$

where Q is the peak discharge in cubic meters per second, and A is the drainage area, in square kilometers. The correlation coefficient (r = 0.894) is significant at the 99.9% confidence level (Students' t test) and the regression is significant at the same level (analysis of variance F-test), but the exponent does not differ from unity at the 10% level. This shows a very good spatial integration of the flood, which is somewhat lost with further increase in drainage area because of rainfall spatial heterogenities. The residuals of the former regression are independent of gullied area, which controls the residuals of the regressions of channel size and June 1981 summer flood peak versus catchment area (for a more detailed analysis see GALLART & CLOTET 1987, and in press). The November 1982 flood peaks are thus

Photo 1: *Massive accumulation of boulders and cobbles in the main channel of a small stream (Llúria stream). Plugging of the channel produced diversion of flow to the right side of the image.*

independent of areas favourable to the production of Hortonian overland flow. They show only a slight relation with hillslope gradient, which can reduce the concentration time and reflect short periods of intense rain within the continuous rainfall event.

The inventory of hillslope phenomena was designed to collect information on both mass movements and new gullies, but the few runoff erosional features which were observed resulted from water flowing from unusual springs in saturated soils and surficial deposits.

The observed changes in the low order streams were deepening by scour of alluvium and widening by bank erosion. The main stream of the badland area of La Barrumba (0.02 km^2 in area), which is normally filled with alluvium, was scoured of most of its alluvial fill in many reaches after the flood. Erosional depths measured in three control cross sections were 17, 24 and 23 cm, in all of which the bedrock was reached (CLOTET et al. 1983).

Changes in the main channels affected not only channel size and shape, but also their paths where valley bottoms are wide enough. Scoured reaches alternate

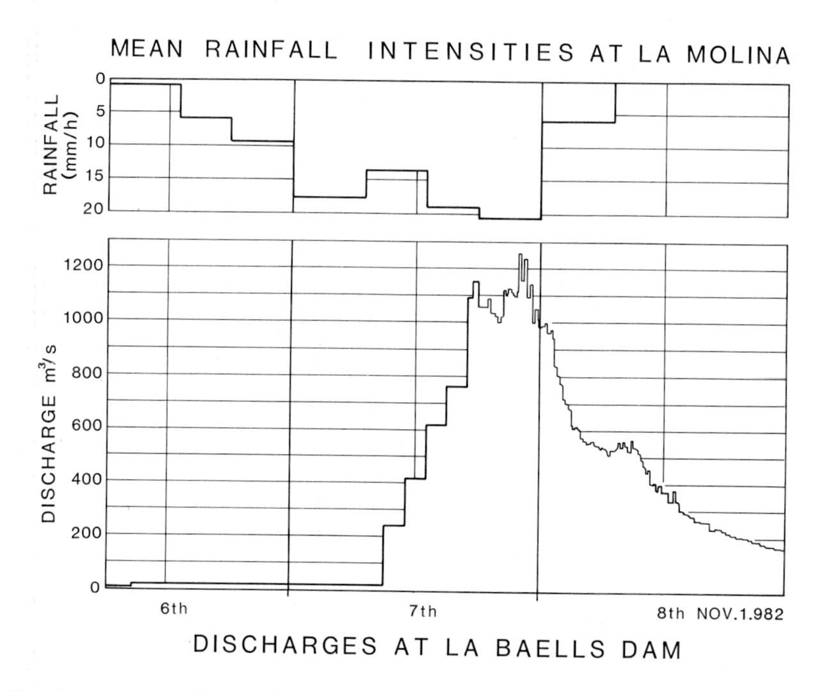

Fig. 3: *Hyetograph at La Molina and hydrograph of Llobregat river at La Baells.*

with massive deposition of cobbles and pebbles which plugged the main channel whilst new channels were eroded by avulsion across the floodplain (photo 1). Most of these changes were assisted by the failure of temporary dams built up by the accumulation of trees and litter. Near to the village of Guardiola, the result of bank erosion and deposition of alluvial bars in the LLobregat river was channel enlargement to about 60 m, while the former channel was less than 20 m wide (detailed maps published in CLOTET & GALLART 1983).

Before the flood, the bedload of the Llobregat river contained 50% of material supplied from coal mining spoil. After the event, this ratio was reduced to 29% as a result of the large amounts of alluvium supplied by bank erosion and scour of the headwater streams (CLOTET et al. 1983b). Unfortunately the flood partly destroyed the gauging station at El Collet, where daily samples of suspended sediment were taken, and no data of suspended sediment during the flood exist.

It is very difficult to estimate the return priod of the flood; since the water stages were much higher than all previous records, but somewhat lower than other 20th. century floods for which the water-level stage has not been recorded. However, for the Segre river the 1982 flood exceeded all previous floods this century (NOVOA 1984), and the above mentioned changes suffered by the Llobregat at Guardiola suggest an unusual duration of the high water stage.

The hydrograph of the flood at the Baells dam shows a broad peak with a relative stabilisation of the discharge at 1100 m^3s^{-1} which lasted about 6 hours (fig.3). This fact suggests that effective rainfall duration exceeded the concentration time, and discharge was thus controlled only by rain intensity, this peak runoff rate is the equivalent of 8 mm per hr, averaged over the entire basin. The unusually long duration of an effective rainfall is thus once more shown to be a characteristic of this extreme event.

4 Mass Movements

A field inventory of the mass movements was designed in order to collect information about the kind, number and extent of such phenomena, their spatial distribution, and relationships with altitude, aspect and gradient of the hillslope, lithology, land use and vegetation cover. Some additional information included former or potential further movement, ocurrence of gullies or badlands, and an estimate of the tendency of the affected area to become a permanently degraded or badland area.

4.1 Types of Movements

As we were concerned with the environmental analysis of a great amount of phenomena rather than with their mechanical analysis, we used field classifying criteria, such as the depth of the affected mass, if it was bedrock or surcicial mantle, and the type of the main and subsequent movements. Our classification was based on standard ones (SHARPE 1938, AVENARD 1962, SCHEIDEGGER 1975, TRICART 1977, VARNES 1978) and a separate type was used for the small failures on artificial slopes in old farming terraces. The types and criteria used are explained as follows:

1. Shallow liquid slides ('debris flow', 'alpine debris flow', 'water blowout', 'coup de cuiller') are shallow (less than 2 metres deep) mass movements of surficial deposits, commonly spoon-shaped. According to the slickensides observed in most of the scars, the movement usually started as a failure, presumably promoted by high pore pressure. There is no major accumulation of material at the foot of the scar, but some scattered in small piles and levees along the track, which evidences a rapid liquid movement of the mass flowing down the hillslope (photo 2).

   This kind of phenomenon has been repeatedly described in the Anglo-American literature, but it is still difficult to classify; the usual term of 'debris flow' or 'mudflow' emphasizes the rough granulometry of the original mass but does not show its origin, nor its speed and water content. According to SCHEIDEGGER (1975) and TRICART (1977), it is necessary to distinguish between slow flows ('tongues' or 'coulées') (photo 5), and rapid flows ('laves torrentielles') (photo 3), rather than their grain size appearance (see also RAPP 1963).

   611 shallow liquid slides were identified, with a total area of 16 ha, resulting in a mean surface of 260 m^2.

2. Slumps ('rotational slide', 'glissement rotational') are wider and deeper failures of bedrock (photo 14) or thick surficial mantle, usually semicircular, but which can also show irregularities resulting from variations in shear strength. the mass usually remains at the foot of the scar or slowly flows as a

Photo 2: *Shallow liquid slide and track of the rapid debris flow issuing from it. On the left the track of another rapid debris flow is visible (Can Lloveres).*

Photo 3: *Spreading of a rapid debris flow issuing from a rather deep shallow liquid slide in surficial deposits, taken about 800 m from the scar (Espinalbet).*

Photo 4: *Slump in Cretaceous marly limestones at El Verd. The upper part of the scar coincides with a forest track. Trees of Pinus sylvestris are about 16 m high; darker trees are Pinus mugo uncinata.*

Photo 5: *A large mass movement near La Coma. The active zone was in Eocene marine clays, but most of the tongue (a slow debris flow) was composed of cobbles derived from Pleistocene surficial deposits. A rapid debris flow like the one on photo 3 preceded the tongue.*

Photo 6: *Surficial slip of soil and surficial deposits over nonmarine Garumnian clays near Fumanya.*

Photo 7: *Agricultural slope failures near La Pobla de Lillet.*

tongue, but some rapid liquid mudflows were generated from cracks (CLOTET & GALLART 1984). Nevertheless, there is a continuous transition between these and the above quoted phenomena. Despite subsequent displacement of most of the mass by fast flows, few slides more than two metres deep were classified as slumps.

195 slumps were observed on the slopes, the affected area being 14 ha, and the mean surface of a single slump 740 m^2. Three great mass movements were identified separately on account of their size, their areas being 25 ha (Torrentsenta), 11 ha (Clot d'Esquers), and 8 ha (La Coma, photo 5). Field evidence of former movements were found in all of them.

3. Surficial slips ('earth slide', 'glissement par paquets') are displacements of the soil or surficial deposits over the bedrock or deeper waste mantle (photo 6). The sliding surface is generally planar, and the displaced mass usually breaks into separate pieces which move independently. There are also transitional phenomena between this kind of movement and the ones quoted above.

   There are 106 surficial slips covering 5.5 ha, with a mean area of 520 m^2.

4. Agricultural slopc failures are small slumps or slides which affect man-made slopes in agricultural terraces (photo 7). They are considered independently because of their small size, great number, and their close dependence on steep artificial slopes. The landslides affecting these areas but which appear to be independent of the artificial slopes were classified in one of the 'natural' groups.

   We found 939 such failures, which cover 11 ha with a mean area of only 120 m^2.

5. Rock falls ('éboulement rocheux') are pieces of bedrock (mainly limestones) detached from cliffs or rocky steps.

   Only 23 such falls were found, covering 1.2 ha with a mean area of 520 m^2.

4.2 Relationships with Hillslope Characteristics

The analysis of the influence of site characteristics on mass movements was made by the construction of contingency tables and analysis by chi-square tests. The first result was the identification of significant effects of site characteristics for most of the contingency tables, this confirms the usefulness of the classification used. the most relevant results are reported below together with the probability of the null hypothesis reported in brackets. The confidence threshold is set at 99% (p=0.01).

1. **Altitude.** Except for rock falls which are more regularly distributed (p=0.025), all kinds of movements show a significant dependence on altitude. The surficial slips have a distribution weakly differenciated from those of slumps (p=0.023), but both differ from shallow liquid slides, which occur in somewhat lower areas. The agricultural slope failures start at 800 m and extend until 1400 m, near the altitudinal boundary for agricultural land use.

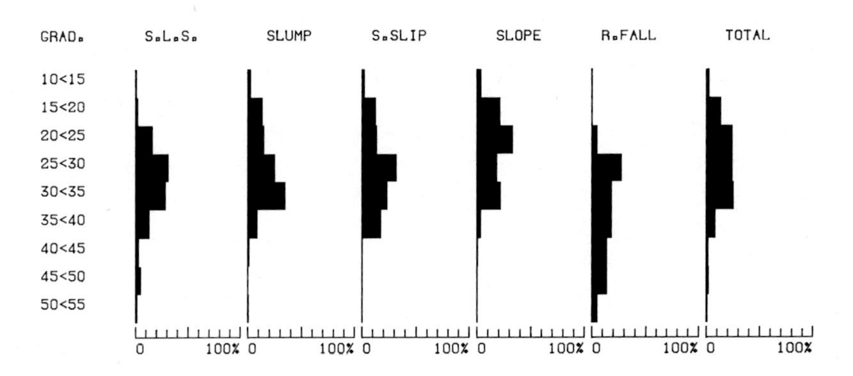

Fig. 4: *Relative frequences of mass movements for changing hillslope gradients. Gradients in degrees.*

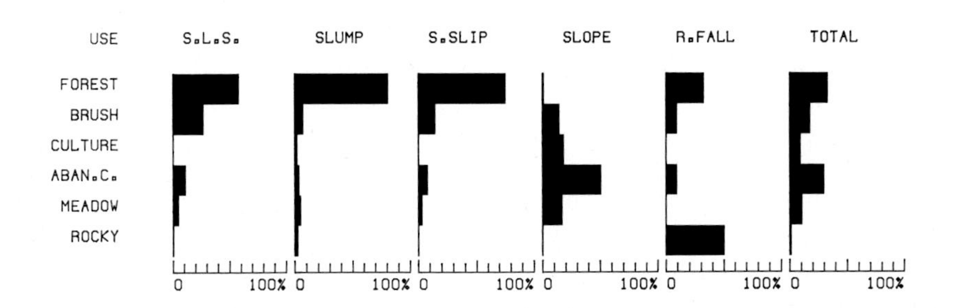

Fig. 5: *Distribution of mass movements for different land uses.*

Raingauge	rainfall mm	total mass movements	non-agricultural mass movements
Vallcebre	342	108	58
La Pobla	340	392	128
Bagà	301	190	23
Cercs	257	17	14
Guilanyà	250	21	16
Borredà	197	1	1
Berga	185	11	0
Tentellatge	122	0	0

Tab. 1: *Rainfall total, number of mass movement phenomena and number of natural mass movements (i.e. excluding agricultural slope failures) in 25 km^2 around each raingauge within the study area.*

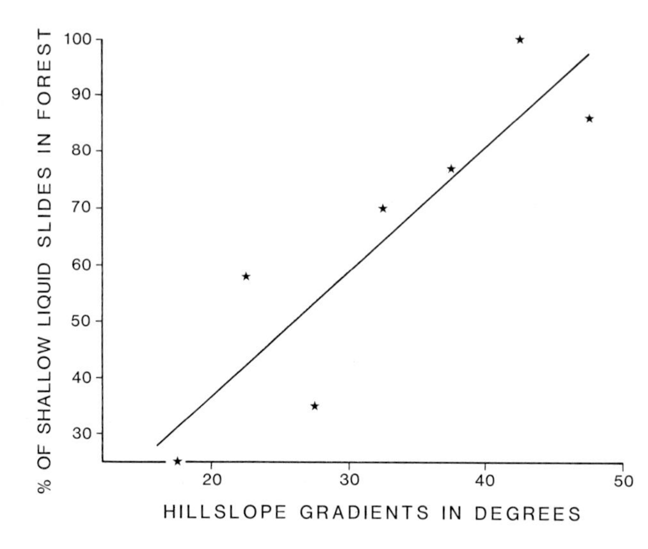

Fig. 6: *Dependence between occurrence of shallow liquid slides in wooded areas and corresponding hillslope gradient.*

The regression equation is $Y = -7.304 + 2.207X$, the correlation coefficient is $r = 0.939$, significant at the 99.9% cinfidence level (Student's t-test), and the regresion is significant at the 98.5% confidence level (analysis of variance F test).

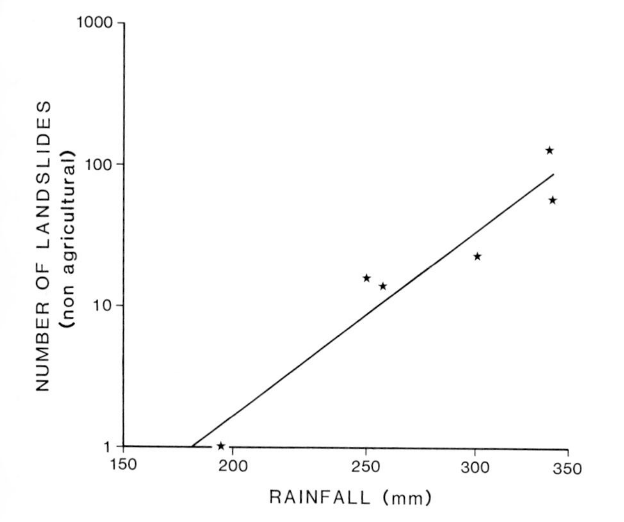

Fig. 7: *Number of mass movements (agricultural slope failures excluded) in 25 km² around every raingauge versus rainfall total for each gauge.*

Both scales are logarithmic and the regresion equation is significant at the 99.6% confidence level. Two raingauges in areas without mass movements collected less than 186 mm.

The most relevant results is the major importance of rock falls (22%) above 1500 m, in zones which only represent 6% of the area in terms of the hypsometric curve. The risk of rock fall is thus 4 times greater than average above this altitude. The rarity of other kinds of mass movements at this height is presumably the result of the hardness of most of the bedrock (mainly limestone) and the thinness of the surficial deposits.

2. **Hillslope gradient (fig.4)**. the gradient values have been measured from a 1:50,000 photogrammetric topographical map with a contour interval of 20 m and therefore correspond to a general gradient rather

than a detailed one. The agricultural slope failures occur on gentler hillslopes (median 25°), but the actual gradient of the affected slopes is usually more than 60°. There is very little difference between slumps and surficial slips, which occur on hillslopes of moderate gradient (median 27°), above a threshold of 15°. Shallow liquid slides and rock falls occur on steeper hillslopes, with a threshold of 20° and with medians of 30° and 35° respectively. On slopes of more than 40° there are only shallow liquid slides (9%) and rock falls (32%).

3. **Bedrock lithology.** This is one of the main controls in most kinds of movements. 37% of failures are in clay, which is the lithology in which all movements other than slope failures and rock falls are most frequent. Rock falls reported on clays are actually in thin limestone beds within a clayey unit, but this reflects an imperfection in the inventory produced by reporting several phenomena in the same form. Shallow liquid slides occur in the same kind of rock as slumps, but surficial slips occur more typically in marls, probably because of a sharper boundary between soil and bedrock.

4. **Aspect of the hillslope.** All kinds of movements other than agricultural slope failures are nearly independent of aspect if the raw values are modified according to the relative frequency of aspects. Agricultural terraces are more frequent in south-facing hillslopes because of the greater suitability of sunny slopes for such land use. Slumps and surficial slips are somewhat more frequent in north-facing hillslopes, whereas rock falls are in south-facing ones. Shallow liquid slides occur on both north and south-facing hillslopes.

5. **Vegetation cover and landuse (fig.5).** The most important association is the high frequency of movement in wooded area. 80% of slumps and 75% of surficial slips occur in woods, which cover only 52% of the studied area. This high frequency of movements is constant across a variety of heights and hillslope gradients. The role of forests in triggering slides under these rainfall conditions is thus suggested.

   58% of shallow liquid slides occur in woods, and if hillslope gradient is also analysed, it appears that the occurrence of such movements in wooded areas increases significantly with increasing gradients (fig.6), the 52% threshold of no influence of forest being at 27°. Trees thus seem to be effective in limiting this kind of movement only at hillslope gradients lower than 27°, but on steeper slopes can trigger such failures. These movements do not occur on rocky slopes, and since steep hillslopes on surficial deposits are typically more wooded than those on rocky ones, the occurrence of shallow liquid slides on wooded steep slopes can be reinforced, independently of any triggering effect of trees.

   The abandoned farmlands suffered 25% of the movements, mainly because 50% of the agricultural slope failures occurred in these areas (65% if these failures in woods and scrub are included as abandoned).

6. **Rainfall spatial distribution.** Maps of the spatial distribution of the number and relative areas covered by mass movements (see fig.2) showed some correlation with the rainfall map. The maps of relative areas of mass movements showed the worse relationship because a few large phenomena disturbed the correlation. In order to analyse the first relationship, the number of mass movements within an area of 25 km^2 around each raingauge was obtained from the original data (tab.1), and the resultant power relationship was computed. The total number of movements shows a good correlation with rainfall amounts, but the regression lacks statistical significance. However, if agricultural slope failures are excluded, the number of mass movements increases with rainfall to the power of 7.4 (fig.7), and the following regression equation is significant at the 99.6% level (analysis of variance F-test):

$$n = 1.39 \times 10^{-17} \times p^{7.424}$$

where n is the number of mass movements and p is the rainfall total in mm. The correlation coefficient is r=0.980, significant at the 99.9% confidence level (Students' t-test). There are two raingauge stations without mass movements which therefore coult not be used in computing the regression, but if the equation is applied to the rainfall collected in these stations, the resultant number is less than unity. Even the null values thus confirm the validity of the regression.

Furthermore, the threshold value of 1 landslide every 25 km^2 occurs at 186 mm of rainfall, not far from 175 mm, the value yielded by the CAINE'S (1980) equation for a 48-hours event. Assuming a normal ditribution of the deviations from the regression, this difference is exceeded 37% of times, and can therefore be accepted.

5 Discussion and Conclusions

Unfortunately, the return period of the 1982 event could not be calculated. In spite of the fact that some of the properties of the storm seem to be relatively frequent in occurrence (e.g. mean rainfall intensities, peak discharges from small catchments), others such as duration of effective rainfall and total amount of water fallen during the whole event seem to represent the highest values recorded this century.

The long duration of high discharges resulted in very significant removal of alluvial deposits from channel beds and banks, which, where valley bottoms were wide enough, produced important changes in the form and path of the main rivers. Rare events seem thus to be of chief importance in the transport of coarse alluvium and in shaping the valley bottoms in mountainous areas (cf. TRICART 1962, HACK & GOODLET 1960, HARVEY 1986).

The high frequency of slumps and surficial slips in wooded areas suggests that forests cover provides a trigger for this kind of movement. Shallow liquid slides seem to be impeded by trees at low hillslope gradient but this role is reversed for gradients steeper than 27°. Many authors agree that forest cover does not provide protection against mass movements during extreme rainfall events, but there are very few suggestions or examples of evidence indicating that forests provide conditions favourable for the initiation of slides. TRICART (1977) suggested that mass movements are frequent in wooded areas because infiltration is higher and the weight of the trees can be decisive in critical conditions. DE PLOEY & CRUZ (1979) emphasized the role of forest root matrix in increasing infiltration and throughflow during very rainy periods; the resultant high hydraulic gradients could reduce shear strength, especially in footslopes. CROZIER (1986) also suggested that high infiltration rates in forest can increase susceptibility to mass movement during heavy rainfall events.

The role of man-made changes in geomorphic processes during the event is then somewhat unclear. On the one hand, most of phenomena were failure of old agricultural slopes, and forest was effective against shallow slides, but on the other hand, deforestation seems to protect hillslopes against rather deep and catastrophic movements. Some reclamation works may be advisable in abandoned agricultural areas strongly affected by slope failure in order to prevent the spreading of gullied areas.

One of the most striking results is the goodness-of-fit in the regression between number of mass movements and rainfall depths. Almost all raingauges are located in clayey depressions and, furthermore, 25 km^2 in such a region is an area large enough to guarantee a low influence of local variations in topographic and lithological conditions. A rapid increase in the number of mass movements with increased rainfall has already been demonstrated elsewhere by GOVI et al. (1982), and EYLES (cited in CROZIER 1986, 175). The relationship demonstrated above can therefore be taken as the result of general physical mechanisms, the fit being enhanced by the weak site control and because the rainfall total is fully representative of such a continuous event.

Acknowledgements

The field work was financed by the Servei Geològic de Catalunya. Meteorologic data are from the Servicio Meteorológico Nacional, and Hydrometric data from the Confederació Hidrogràfica del Pinineu Oriental. We acknowledge T. Dunne, A.M. Harvey, and M. Sorriso-Valvo for their critical review of the manuscript. We are also very grateful to the anonymous referees from CATENA for their comments and English style improvements.

References

AVENARD, J.M. (1962): La solifluxion ou quelques méthodes de mécanique des sols appliquées au probleme géomorphologique des versants. Trav. Lab. Geogr. Phys. Centre de Geogr. App. Univ. Strasbourg, **1**, 164 pp.

CAINE, N. (1980): the rainfall intensity-duration control of shallow landslides and debris flows. Geografiska Annaler **62A**, 23–27.

CLOTET, N. & GALLART, F. (1983): Dinàmica a la conca de lAlt Llobregat. In: Efectes geomorfològics dels aiguats de novembre de 1982, informes **1**, 48–113. Pub. Servei Geològic de Catalunya.

CLOTET, N. & GALLART, F. (1984): El deslizamiento de La Coma (Solsonès, Catalunya) de Novembre de 1982. Inestabilidad de Laderas en el Pirineo, Ponencias y comunicaciones. Barcelona, 1.6.1.–1.6.14.

CLOTET, N., GALLART, F. & CALVET, J. (1983): Estudio de la dinámica de un sector de badlands en Vallcebre (Prepirineo catalán): Actas de la II Reunión del Grupo Español de Geología Ambiental y Ordenación del Territorio. 4.20–4.39. Lleida.

CLOTET, N. & GALLART, F. (1986): Sediment yield in a mountainous basin under high

Mediterranean climate. Zeitschrift für Geomorphologie Supp. Bd. **60**, 205–216.

CLOTET, N. & GALLART, F. (in press): Inventari de les degradacions de vessants originades pels aiguats de novembre de 1982 a les altes conques del Llobregat i Cardener. Servei Geològic de Catalunya.

CROZIER, M.J. (1986): Landslides. Causes, consequences and environment. Croom Helm, London, 252 pp.

DE PLOEY, J. & CRUZ, O. (1979): Landslides in the Serra do Mar, Brazil. CATENA **6**, 111–122.

GALLART, F. & CLOTET, N. (in press): Recherches sur le modélé des lits dans une zone de montagne méditerranéenne. Proposition dun indice de portée relative des lits torrentiels fondé sur leur géometrie. Recherches Géographiques à Strasbourg.

GALLART, F. & CLOTET, N. (1987): Channel hydraulics as a method for hydrological analysis of uninstrumented catchments. In: GARDINER, V. (Ed.), International Geomorphology 1986, part I. Wiley, Chichester, 711–721.

GOVI, M., SORZANA, P.F. & TROPEANO, D. (1982): Landslide mapping as a evidence of extreme regional events. Studia geomorphologica Carpatho-Balcanica **15**, 81–97.

HACK, J.T, & GOODLET, J.C. (1960): Geomorphology and forest ecology of a mountain region in the Central Appalachians. U.S. Geological Survey Prof. Paper, **347**, 66 p.

HARVEY, A.M. (1986): Geomorphic effects of a 100 year storm in the Howgill Fells, Northwest England. Zeitschrift für Geomorphologie N.F. **30, 1**, 71–91.

MARTIN-VIDE, J. & RASO, J.M. (1983): Marc atmosfèric. In: Efectes geomorfològics dels aiguats de novembre de 1982. Infomes **1**, 14–31. Pub. Servei Geològic de Catalunya.

NOVOA, M. (1084): Precipitaciones y avenidas extraordinarias en Catalunya. Inestabilidad de laderas en el Pirineo: Ponencias y comunicaciones Universitat Politècnica de Barcelona, 1.1.1.–1.1.15.

RAPP, A. (1963): The debris slides at Ulvadal, Western Norway. An example of catastrophic slope processes in Scandinavia. Nach. Akad. Wiss. Göttingen, II Math. Phys. Klasse, **13**, 195–210.

SCHEIDEGGER, A.E. (1975): Physical aspects of natural catastrophes. Elsevier, Amsterdam, 289 pp.

SHARPE, C.F.S. (1938): Landslides and related phenomena. Columbia Un. Press, New York, 137 pp.

TRICART, J. (1962): Les transports solides du Vidourle lors de la crue d'automne 1958. Actes du 86 Congrès des Soc. Savantes, Sect. de Géographie, 453–535.

TRICART, J. (1977): Précis de Géomorphologie, Tome II: Géomorphologie Dynamique Générale S.E.D.E.S. Paris, 345 pp.

VARNES, D.J. (1978): Slope movements types and processes. Landslides. Analysis and control. Spec. Report **176**, Transportation Res. Board, Nat. Acad. Sc. Washington, 11–33.

Address of authors:
F. Gallart, N. Clotet-Perarnau
Institut Jaume Almera
Ap 30102, 08028
Barcelona, Spain

Jan de Ploey (Ed.)

RAINFALL SIMULATION, RUNOFF and SOIL EROSION

CATENA SUPPLEMENT 4, 1983

Price: DM 120,-

ISSN 0722–0723 ISBN 3–923381–03–4

This CATENA-Supplement may be an illustration of present-day efforts made by geomorphologists to promote soil erosion studies by refined methods and new conceptual approaches. On one side it is clear that we still need much more information about erosion systems which are characteristic for specific geographical areas and ecological units. With respect to this objective the reader will find in this volume an important contribution to the knowledge of active soil erosion, especially in typical sites in the Mediterranean belt, where soil degradation is very acute. On the other hand a set of papers is presented which enlighten the important role of laboratory research in the fundamental parametric investigation of processes, i.e. erosion by rain. This is in line with the progressing integration of field and laboratory studies, which is stimulated by more frequent feed-back operations. Finally we want to draw attention to the work of a restricted number of authors who are engaged in the difficult elaboration of pure theoretical models which may pollinate empirical research, by providing new concepts to be tested. Therefore, the fairly extensive publication of two papers by CULLING on soil creep mechanisms, whereby the basic force-resistance problem of erosion is discussed at the level of the individual particles.

All the other contributions are focused mainly on the processes of erosion by rain. The use of rainfall simulators is very common nowadays. But investigators are not always able to produce full fall velocity of waterdrops. EPEMA & RIEZEBOS give complementary information on the erosivity of simulators with restricted fall heights. MOEYERSONS discusses splash erosion under oblique rain, produced with his newly-built S.T.O.R.M–1 simulator This important contribution may stimulate further investigations on the nearly unknown effects of oblique rain. BRYAN & DE PLOEY examined the comparability of erodibility measurements in two laboratories with different experimental set-ups They obtained a similar gross ranking of Canadian and Belgian topsoils.

Both saturation overland flow and subsurface flow are important runoff sources the rainforests of northeastern Queensland. Interesting, there, is the correlation betwe colour and hydraulic conductivity observed by BONELL, GILMOUR & CASSELLS. R generation was also a main topic of IMESON's research in northern Morocco, stressi mechanisms of surface crusting on clayish topsoils.

For southeastern Spain THORNES & GILMAN discuss the applicability of er models based on fairly simple equations of the "Musgrave-type" After Richter (Ger and Vogt (France) it is TROPEANO who completes the image of erosion hazards in Eur vineyards. He shows that denudation is at the minimum in old vineyards, cultivate manual tools only. Also in Italy VAN ASCH collected important data about splash erosi rainwash on Calabrian soils. He points out a fundamental distinction between tran limited and detachment-limited erosion rates on cultivated fields and fallow land. representative first order catchment in Central-Java VAN DER LINDEN comment trasting denudation rates derived from erosion plot data and river load measurements too, on some slopes, detachment-limited erosion seems to occur.

The effects of oblique rain, time-dependent phenomena such as crusting and generation, detachment-limited and transport-limited erosion including colluvial depo are all aspects of single rainstorms and short rainy periods for which particular, pre models have to be built. Moreover, it is argued that flume experiments may be an eco way to establish gross erodibility classifications. The present volume may give an imp further investigations and to the evaluation of the proposed conclusions and suggesti

Jan de Ploey

CATENA SUPPLEMENT 13 p. 97–107 Braunschweig 1988

REGIONAL EROSION: RATES AND SCALE PROBLEMS IN THE BUONAMICO BASIN, CALABRIA

P. **Ergenzinger**, Berlin

Summary

The Buanomico basin in Calabria is a mountain area with Quaternary uplift of more than 1000 m and marked spatial and temporal variability of erosion.

Mass movements on the slopes are the main erosional features. Most of the slid material is stored on the lower part of the slopes. Some is transported, especially during storm runoff, in braided river systems. The valley floor is a second main store of this material.

The long recurrence interval between extreme precipitations means that observation periods of more than 25 years are necessary for the medium-term prediction of erosion rates. To determine the regional variability of mass movement the area to be investigated should be bigger than 20 km^2.

Quaternary longterm erosion is extremely dependent on tectonic history. Uplift rates and erosion rates are of the same order.

Resumen

La cuenca de Buonamico forma parte del Aspromonte, en Calabria, y es la montaña más meridional de la península italiana. El punto más alto de la cuanc, el Montalto (1956 m) está a solamente 26 km al norte del mar Iónico. El levantamiento Cuaternario del Aspromonte es superior a los 1000 m y esta acción tectónica tan fuerte ha dado lugar a una también extremada erosión en las vertientes y en los fondos de valles.

Hay una gran variabilidad espacial en la movilización de materiales sólidos en vertientes y canales. Las elevadas tasas de erosión que se producen en las vertientes a causa de los movimientos en masa condicionan el desarrollo de los fondos de valle. La principal fuente de carga fluvial son las vertientes empinadas de la parte media y alta de los valles. No hay una relación simple entre precipitaciones máximas, erosión y producción de sedimento. La producción de sedimento al pie de los releives montañosos está influído por la acumulación de material procedente de deslizamientos en la parte baja de las vertientes y en el fondo de los ríos con canales trenzados. La principal acumulación de sedimento tiene lugar en el amplio (>200 m) fondo aluvial de la parte baja de la cuenca del Buonamico. La acumulación y la erosión en este fondo aluvial fueron medidas geodésicament durante una década y determinada la tasa de erosión, la producción de sedimento y el transporte de

ISSN 0722-0723
ISBN 3-923381-13-1

3-923381-13-1/88/5011851/US$ 2.00 + 0.25

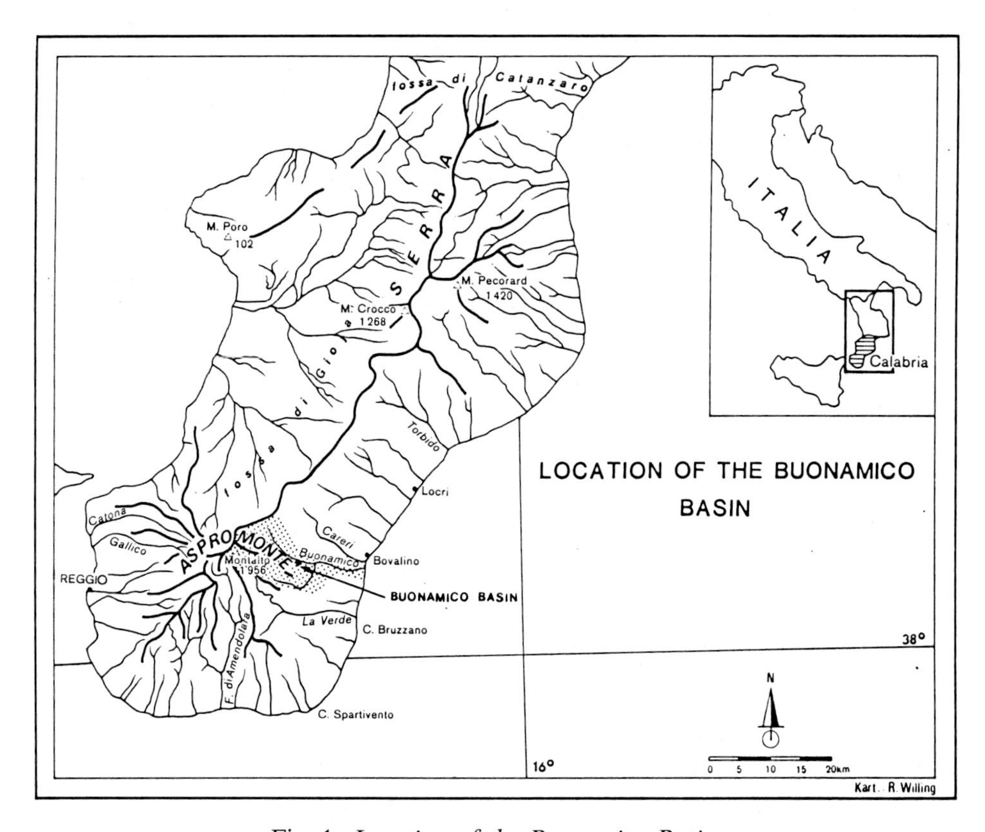

Fig. 1: *Location of the Buonamico Basin.*

sedimento del período 1971–1980.

Existe también una gran variabilidad temporal en la erosión. La capacidad para la erosión y el transporte fluctúa entre diferentes eventos en más del factor 1000. En 1971–72 tuvo lugar el último evento "grande". La comparación de fotografías aéreas y las investigaciones sobre movimientos en masa muestra que estos sucesos tienen períodos de recurrencia cercanos a los 25 años. El estudio del balance de sedimentos del rio La Verde dió como resultado tasas de remoción de sedimento de más de 1000 mm/ 100 años. Estas tasas de erosión tan altas hay que situarlas en el contexto del terremoto de Messina-Reggio de 1908.

Si el total de la erosión Cuaternaria se calcula mediente la determinación del volumen erosionade en los valles la erosión a largo plazo no se halla equitativament distribuida en la cuenca. La máxima erosión tienen lugar cerca de la desembocadura y en el tercio superior del perfil longitudinal.

1 Introduction

Regional erosion, the output of solid material from a specific basin, can be dealt with at very different scales of time and space (GREGORY & WALLING 1973). Taking the Buonamico basin, Calabria, as an example, we can deal with a specific slope of a valley in 1971, or with the total erosion of the basin during the Quaternary era. For the Quaternary we derive an overall erosion rate with the whole basin as the source area, whereas for short-term microscale erosion the concept of variable source areas must be used. In the case of the Buonamico the lower limits of space and time are of special interest. When these limits and the frequency and magnitude (WOLMAN & MILLER 1960) of the important processes are known, a minimum observation time and area of regional erosion may be determined.

2 Regional Setting

The Buonamico basin in Calabria, Southern Italy, occupies a total area of 146 km^2. It is part of the crystalline Aspromonte, the southernmost mountain of the Italian boot (fig.1). The highest point of the basin, the Montalto (1956 m), is situated only 26 km from the Ionian Sea. Quaternary uplift of the Aspromonte exceeds 1000 m. This extreme tectonic mobility has led to extreme erosion of slopes and valley bottoms. The first detailed geomorphological survey was made by LEMBKE (1931).

3 Long-term Basin Erosion

By reconstructing the former surface at the boundary between Pliocene/Quaternary GÖRLER & UCHDORF (1980) calculated the difference in volume between the present landscape and the former one. The intensive tectonic uplift was accompanied by a corresponding regional erosion. But even the long-term regional erosion is not evenly distributed. Close to the divides there are some relicts of the former landscapes (former nearshore accumulations, known as "Calabriano") with almost no erosion. Maximum erosion rates are typical of the lower end of the upper third of the longitudinal profile of the Buonamico. Towards the Ionian Sea the erosion rates fall again. Quaternary long-term erosion depends primarily on tectonic uplift. Uplift rates and erosion rates tend to be of the same order (0.5 mm per year, Quaternary = 2×10^6 years).

In the lower Buonamico valley there was an intensive accumulation during the last 15000 years (COTECCHIA & MELIDORO 1974). The volume of trapped bedload was determined by ERGENZINGER & GEHRENKEMPER and by ITALPROS independently at $300 \times 10^6 m^3$. Redistributed onto the basin area this is 2 m per 15000 years, or roughly 0.14 mm per year. However, over the same period total erosion rates approximately double this figure since only the coarse material was trapped in the valley and the ratio between bedload and washload in close to 0.5.

4 Erosion Rates over 100-Year Periods

Reliable data on erosion over several centuries are extremely rare. One example was presented by IBBEKIN & RUMOHR (in ERGENZINGER et al. 1978) for the La Verde basin, a watershed next to the Buonamico basin. By comparing the shelf topography shown in a nautical map of 1876 with the present-day situation they derive a sediment delivery rate for the La Verde basin of close to 1000 mm per 100 years. This is

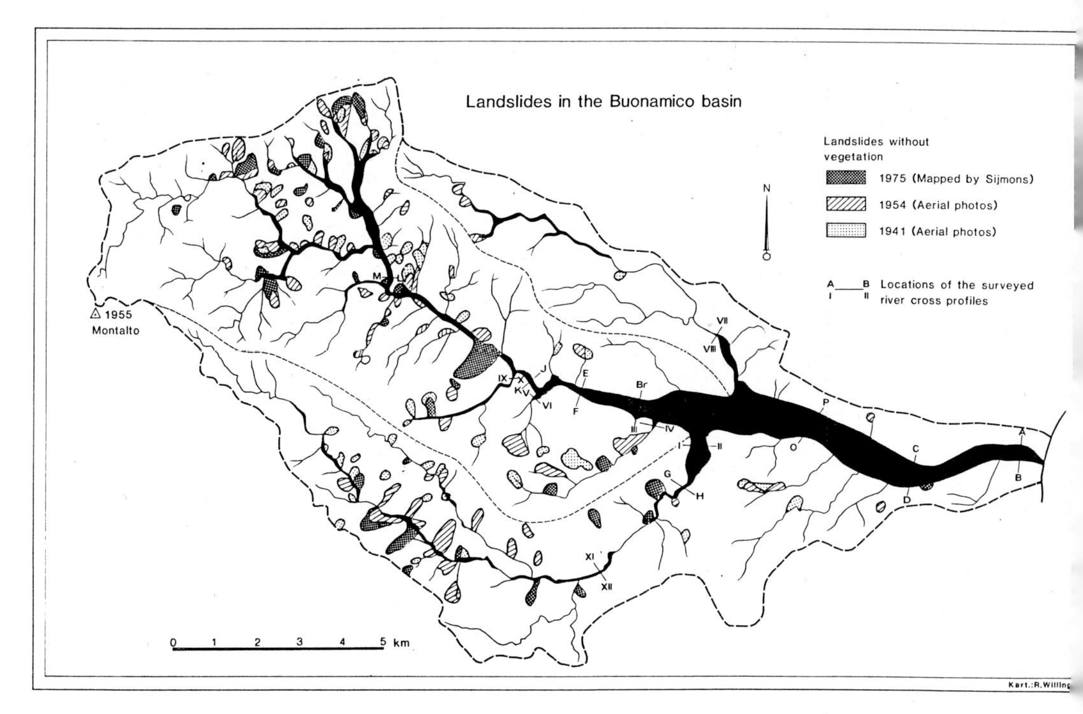

Fig. 2: *Landslides in the Buonamico Basin.*

an extreme example, but in reality this value is even higher because only a certain proportion of the transported sediment would have been deposited on the shelf. If the old map is reliable there must have been one or several extreme events in the La Verde basin. The authors propose a connection between the extreme erosional events and the Messina-Reggio earthquake of 1908. Triggered by the earthquakes there must have been a great number of mass movements (known in Italy as "frane"). This was shown for the earthquake of 1783 by COTECCHIA, MELIDORO & TRAVAGLINI (1969). The recurrence interval of comparable earthquakes is only 250 to 300 years.

5 Short-term Regional Erosion

For the periods from 1954 to 1980 and especially from 1971 to 1980 some observations concerning erosion, erosion rates, sediment transport and sediment delivery were made by OBENAUF & ERGENZINGER. By comparing three sets of air photos, mass movement and related slope erosion during the following time intervals were determined for the mountainous parts of the Buonamico basin (120 km^2) (fig.2): Mass movements affected 2.1, 4.3 and 3.2 km^2 prior to 1941, from 1941 to 1954, and from 1954 to 1980. These mass movements occurred in connection with catastrophic flood events in 1933, 1951, 1958 and 1972 (fig.3).

The volume of the big frane can be determined by estimating an average thick-

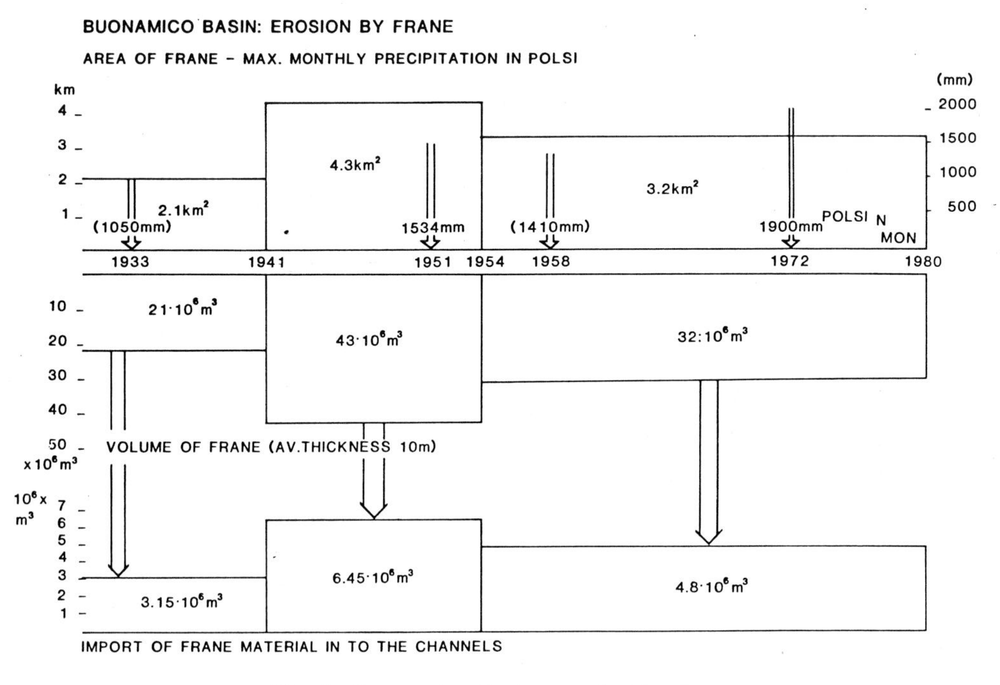

Fig. 3: *Buonamico Basin: erosion by frane.*

ness of 10 to 30 metres. Even with a conservative estimate of only 10 metres, close to 100 × $16^6 m^3$ were mobilized during the last 50 years. The distribution of slopes with big mass movements (fig.2) corresponds to the occurrence of long slopes at the lower and middle part of the basin in Aspromonte. The uppermost part of the basin was much less affected by frane. Most of the material remained on the slopes. On average, only 15% of the volume moved by the frane was afterwards transported into the river channels. Based on these assumptions the resulting erosion during the last 50 years was 15 × $10^6 m^3$ or, close to 2.5 mm per year over the basin as a whole. In reality the erosion was not evenly distributed but concentrated at several point sources, with local inputs of slope material during specific times after the landslide events.

On the slopes affected by landslides since 1933 (9.6 km^2) there has been slope erosion of close to 1.6 m km^{-2}. But even this erosion is not distributed uniformly in space and time. This will be demonstrated with reference to the Costantino landslide.

December 1972 was very wet. In Polsi (786 m) in the upper Buonamico basin there was about 1850 mm of precipitation. The maximum daily precipitation (about 390 mm) occurred on 3 January 1973 (GUERRICCHIO & MELIDORO 1973). The Constantino landslide occurred at the end of this extreme event during the night of January 3/4. According to topographic and geological surveys the volume of the frane was about 21 × $16^6 m^3$. The landslide crossed the middle Buonamico valley and dammed the river. Up to 3.6 × $10^6 m^3$ of water

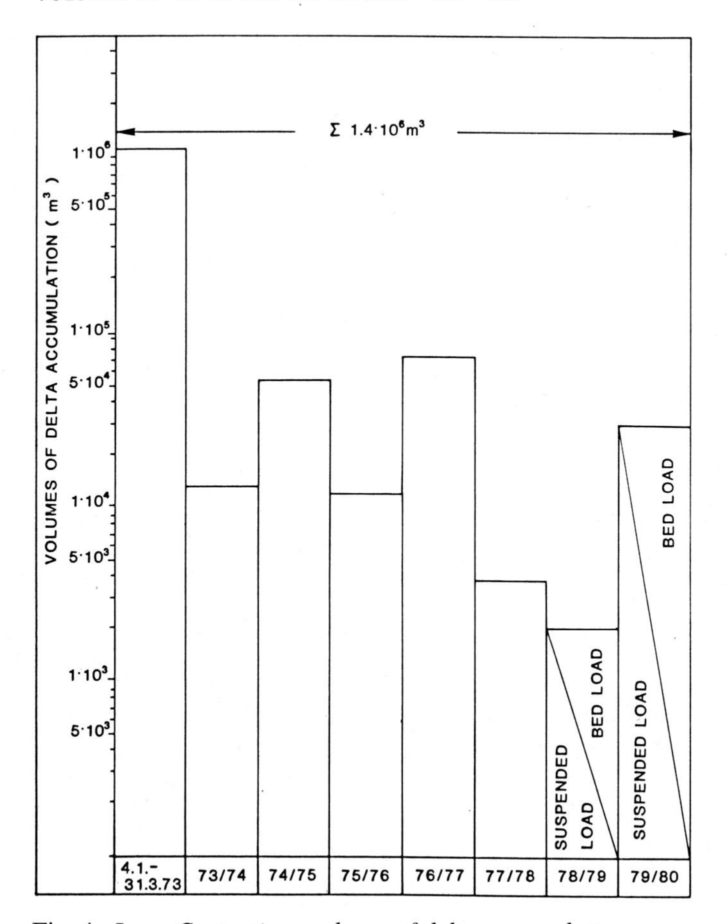

Fig. 4: *Lago Costantino: volume of delta accumulation per year.*

accumulated. After some rain showers the lake water filtered through the landslide dam and eroded a canyon close to 40 m deep during approximately 24 hours on 4 February 1973. The outflow of more than $2.6 \times 10^6 m^3$ of water lowered the lake level to about 20 metres. The material eroded by the canyon comprised $1.2 \times 10^6 m^3$. During the following years this remarkable rate of erosion never occurred again in the new canyon. The longitudinal profile was stabilized by huge boulders close to the outlet of the lake. The canyon erosion can best be described by a negative e-function, with a short period of intensive erosion and canyon formation, followed by a much longer period of minor erosion without major changes to the morphology of the canyons: a process with a long relaxation time (THORNES 1979).

After 1974 intensive accumulation took place below the frane, for a former secondary valley was re-created in the

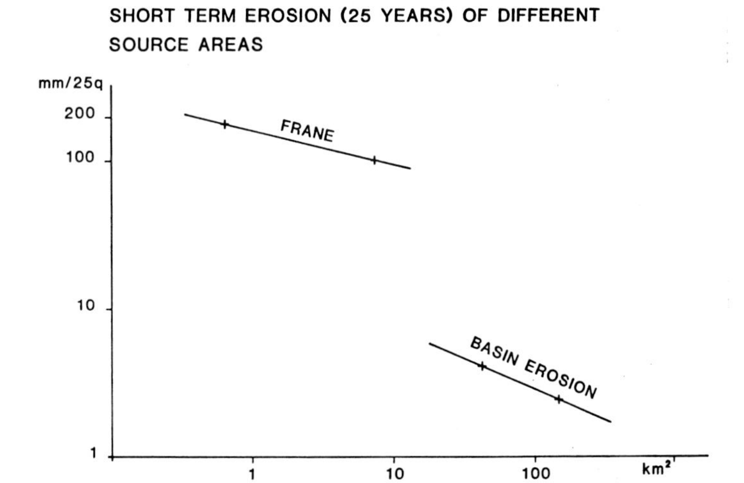

Fig. 5: *Short-term erosion (25 years) of different soucre areas.*

central part of the landslide. The eroded material totalled 1.8 × 10^6m^3 for the period from 1974 to 1980. The erosion of the small and deeply incised secondary valley diminished on the metamorphic gneiss bedrock close to the base of the landslide. The landslide material eroded by the canyon and the secondary valley totalled 3.2 × 10^6m^3 between 1973 and 1980. Related to the landslide area of 0.73 km^2 this is on average more than 4 metres in 8 years.

For the same period it is possible to compile a sedimentary budget for the upper Buonamico basin above Lago Costantino (41 km^2). At the end of the flood event of 1972/73 at the inflow to the new Lago Costantino a delta was created with a volume of 1.1 × 10^6m^3 (fig.4). During the following years up to 1980 the solid material in the delta and the lake basin amounted to 1.4 × 10^6m^3 plus 0.6 × 10^6m^3. The trapping efficiency of the lake is 95% of the total incoming suspended load. The minimum accumulation of solid material in the lake during one winter was 17000 m^3, the secondary maximum was close to 70000 m^3 per winter. During the second half of the winter of 1972/73 more than double the volume was transported out of the upper Buonamico basin compared to the following seven years. The total erosion volume during the winter of 1972/73 in the upper basin must have been higher than 3 × 10^6m^3 or more than 7 centimetres compared to the 1.5 centimetres for the rest of the period. But again more than 70% of this material came from landslide areas and material was transported only during floods. The transportation efficiency of the Aspromonte rivers is indicated by the measurements of suspended load at Butramo (draining 40 km^2) in February 1975 with 25000 t or close to 40000 m^3. This figure approximates to 1 mm of erosion in one month by two flood waves and relates to suspended material alone.

During the winter of 1972/73 erosion and accumulation on the river bottom of the Buonamico system resulted in

+	2.4	×	$10^6 m^3$
-	0.9	×	$10^6 m^3$
+	1.5	×	$10^6 m^3$

(compare ERGENZINGER et al. 1975). Due to the erosion of the Costantino landslide the accumulation on the river bottom up to 1980 was close to $3.5 \times 10^6 m^3$. The output into the Ionian Sea is estimated at more than $1.1 \times 10^6 m^3$ for the same period. The recovery time to erode and transport the surplus material from the last major event is therefore 20–25 years, and is comparable with some observations of WOLMAN & GERSON (1978) on semiarid rivers.

6 Results

There is a high of scatter of data for regional erosion in the Buonamico basin. As long as the time base for the extrapolation is not determined and the related area under consideration has only an upper limit (the total area of the basin) and no lower limit, many different values for regional erosion will result.

In the case of fluvial erosion, the timescale for regional erosion must be determined by the recurrence interval of big floods. For regional fluvial erosion and transportation, the threshold events are not bankfull rivers but flooding rivers with inundated high terraces or, as in Calabria, inundated braided river bottoms. At the Buonamico the flows at these stages have not been measured. However, daily rainfall has been recorded at Polsi for the past 50 years. Every 20 to 25 years there has been catastrophic precipitation with high floods and several frane movements. A 25-year interval is therefore appropriate as the observation period for regional erosion.

Tab.1 shows the short-term erosion rates of several areas based on this 25-year interval. Thus the difficulties with short-term local exponential erosion are avoided.

The erosion rate on slopes affected by major mass movements is more than 100 times higher than the erosion rate in the related river basin. The main temporary receivers of coarse material are the lower parts of the slopes and the river bottom. On the Buonamico, regional and fluvial erosion rates may be similar over periods of 25 years, 15000 years, and 2×10^6 years. However, they vary considerably for different sources and different areas.

This is typical of river basins with a high rate of coarse material production whereas for basins with a high rate of suspended load LEOPOLD et al. (1964) stated that "a large proportion of the work is performed by relatively frequent events of moderate magnitude".

The magnitude and frequency of sediment yield by the river Buonamico and on the slopes are shown in fig.6. The magnitudes range from 1000 to $10 \cdot 10^6 m^3 yr^{-1}$. The frequencies of sediment output from the slopes are biased: the highest frequency close to 70% is related to low magnitude and, vice versa, for the high magnitude there is a low frequency. The sediment yield transported by the river is more unimodal with the highest frequency around a sediment yield of 10000 $m^3 yr^{-1}$. But again the range of magnitude is large. The importance of the low frequency-high magnitude events can be demonstrated by the calculation of the erosional work done by the rivers and by the mass movements on the slopes.

The results of multiplying different magnitudes by the related frequencies are depicted in fig.7. The maximum for river work is close to the magnitude of $10^7 m^3 yr^{-1}$ and the same is true for the

	area (km^2)		volume ($10^6 m^3$)	$\frac{\text{volume}}{\text{area} \times 25\text{years}}$ (mm/25 years)
SLOPE EROSION				
single frana Costantino	0.72		3.2	178
Buonamico frane	7.5		11.3	100
BASIN EROSION				
upper Buonamico basin	41		3.0	4.7
		0.6 × 3*	1.8	
			4.8	
total Buonamico basin	146		1.5	2.4
			3.5	
			1.1	
		0.9 × 3*	2.7	
			8.8	

* Extrapolation of the measured data 1972–1980 to 25 years

Tab. 1: *Short-term erosion rates for different areas.*

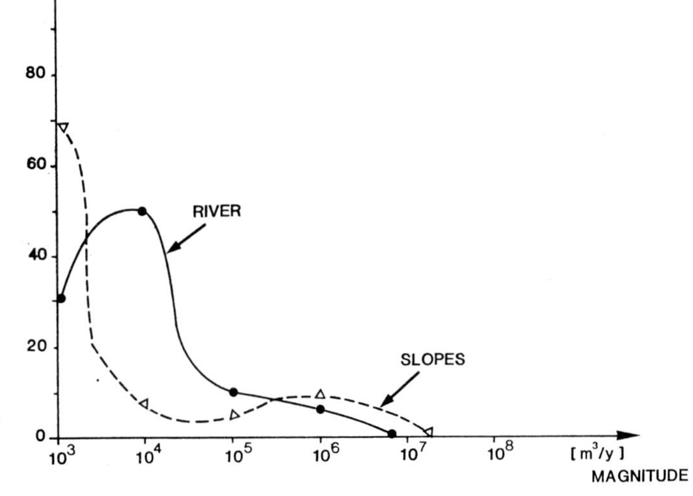

Fig. 6: *Buonamico Basin: sediment yield by fluvial transport and mass movement.*

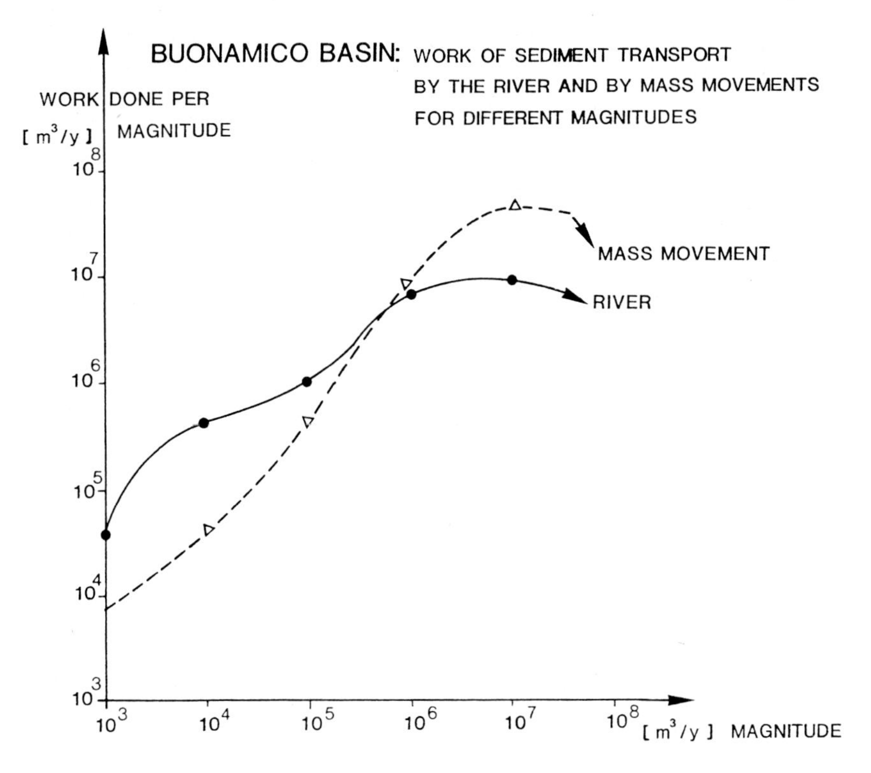

Fig. 7: *Buonamico Basin: work of sediment transport by the river and by mass movements for different magnitudes.*

mass movements. Both figures stress the importance of the low frequency-high magnitude events. This result confirms the remarks of LEOPOLD et al. (1964): for basins with braided rivers and a high rate of coarse bedload transport the less frequent events are more important. Under these conditions the minimum observation time is close to 25 years and the basin should be so large that there is at least a high probability of a large frana. The minimum area for the determination of regional erosion in the southern Aspromonte is: total mountainous area (close to 120 km^2) divided by average area of frana (about 3 km^2) = 40 km^2. The information collected in the Costantino basin is representative as regards the aerial dimension, but the observation time is still far too short. Furthermore, these extrapolations do not include the importance of the super event: a combination of earthquake and intensive precipitation.

References

COTECCHIA, V., MELIDORO, G. & TRAVAGLINI, G. (1969): I movimenti franosi e gli sconvolgimenti della rete idrografica prodotti in Calabria dal terremoto del 1783. Geologia Applicata e Idrogeologia **IV**.

COTECCHIA, V. & MELIDORO, G. (1974): Some principal Geological Aspects of the landslides of southern Italy. Bulletin of the Int. Assoc. of Engineering, Geology **9**, 23–32.

ERGENZINGER, P., GÖRLER, K., IBBEKEN, H., OBENAUF, P. & RUMOHR, J. (1978):

Calabrian Arc and Ionian Sea: Vertical Movements, Erosional and Sedimentary Balance. In: Alps, Apennines, Hellenides. Inter-Union Committee on Geodynamics, Scientific Report **38**, 359–373, Stuttgart.

ERGENZINGER, P., OBENAUF, K.P. & SIJMONS, K. (1975): Erster Versuch einer Abschätzung von Erosions- und Akkumulationsbeträgen einer Torrente Kalabriens. Würzburger Geographische Abhandlungen **43**, 174–186.

GÖRLER, K. & UCHDORF, B. (1980): Zur quantitativen Erfassung von Erosion und Denudation — Beispiele aus Süditalien. Berliner Geowissenschaftliche Abhandlungen, Reihe A / Band **20**, 121–136.

GREGORY, K.J. & WALLING, D.E. (1973): Drainage Basin form and process — a geomorphological approach. (Arnold), London.

GUERRICCHIO, A. & MELIDORO, G. (1973): Segni premonitori e collani della grandi frane nelle metamorfiti della valle della Fiumara Buonamico (Aspromonte, Calabria). Geologia Applicata e Idrogeologia **VIII**, 315–337.

ITALPROS (Ed.) (1983): Studi ed indagini geologichi, geofisiche, idrologiche et idrogeologiche nei bacini alluvionali delle fiumara comprese tra lo Stilaro et il T. Melito pr la corretta valutazione delle risorge idriche dei subalvei nelle stagione di Magra e per la eventuale ricarica di serbatoi sotterranei alluvionali Cassa per il Mezzogiorno, Reggio Calabria.

LEMBKE, H. (1931): Beiträge zur Geomorphologie des Aspromonte (Kalabrien). Zeitschrift f. Geomorphologie, hbfVI , 58–112.

LEOPOLD, L.B., WOLMAN, M.G. & MILLER, J.P. (1964): Fluvial Processes in Geomorphology. (Freeman), San Francisco. 522 pp.

THORNES, J.B. (1979): Processes and interrelationships; rates and changes. In: EMBLETON, C. & THORNES, J.B. (Eds.), Process in Geomorphology. 378–387, (Arnold), London.

WOLMAN, M.G. & MILLER, J.P. (1960): Magnitude and frequency of forces in geomorphic processes. Journal of Geology, **68**, 54–74.

WOLMAN, M.G. & GERSON, R. (1978): Relative scales of time and effectiveness of climate in watershed geomorphology. Earth Surface Processes, **3**, 189–208.

Address of author:
P. Ergenzinger
Freie Universität Berlin
Fachbereich Geowissenschaften
Institut für Physische Geographie
Grunewaldstraße 35
D 1000 Berlin 41
F.R.G.

Asher P. Schick (Ed.):

CHANNEL PROCESSES

WATER, SEDIMENT,
CATCHMENT CONTROLS

CATENA SUPPLEMENT 5, 1984

Price DM 110,—

ISSN 0722–0723 / ISBN 3–923381–04–2

PREFACE

Two decades ago, the publication of 'Fluvial Processes in Geomorphology' brought to maturity a new field in the earth sciences. This field – deeply rooted in geography and geology and incorporating many aspects of hydrology, climatology, and pedology – is well served by the forum provided by CATENA. Much progress has been accomplished in fluvial geomorphology during those twenty years, but the highly complex and delicate relationships between channel processes and catchment controls still raise intriguing problems. Concepts dealing with thresholds and systems, and modern tools such as remote sensing and sophisticated tracing, have not decisively resolved the simple but elusive dual problem: how does the catchment shape the stream channel and valley to its form, and why? And: how does the channel transmit its influence upstream in order to make the catchment what it is?

Partial solutions, in a regional or thematic sense, are common and important. In addition to contributing a building block to the study of fluvial geomorphology, they also produce a number of new questions. The consequent proliferation of research topics characterises this collection of papers. The basic tool of geomorphological interpretation – the magnitude, frequency, and mechanism of sediment and water conveyance – is a prime focus of interest. Increasingly important in this context in recent years is the role of human interference natural fluviomorphic process systems. Effects of drainage ditching, transport of pesticides absorbed in fluvial sediment, and the flushing of nutrients are some of the Manconditioned aspects mentioned in this volume. Other contributions deal with the intricate balance, especially in extreme climatic zones, between physical process generalities and macroregional morphoclimatic influences.

The contributions of PICKUP and of PICKUP & WARNER represent two of the very few detailed quantitative geomorphological analyses of very humid tropical catchments. The 8 to 10 m mean annual rainfall in the equatorial mountain areas studied combines with effective landsliding to produce extremely high denudation rates. However, many aspects of channel behaviour are similar to those of temperate rivers. Particularly interesting are the relationships derived between channel characteristics, perimeter sediment and bedload transport.

Several small ephemeral and intermittent streams in Ohio studied by THARP, although variable in catchment area and in peak discharge, have a similar competence; while sorting increases downstream, the coarsest sizes tend to remain constant. Sorting of fluvial sediment, though on a much longer time scale and in an arid climate, also plays an important role in the contributi MAYER, GERSON & BULL. They find that modern channel sediment size bits the most rapid downstream decrease in mean particle size, while Pleist deposits show the least rapid decrease and are consistently finer than yc deposits. The difference is attributed to climatic change and a predictive thereto is presented.

HASSAN, SCHICK & LARONNE describe a new method for the ma tracing of large bedload particles capable of detecting tagged particles redep by floods up to several decimetres below the channel bed surface. Their m may considerably enhance the value of numerous experiments with p pebbles, previously reported or currently in progress. Suspended sediment subject of the paper by CARLING. He experiments with sampling gravel-b flashy streams by two methods, and concludes that pump-sampling and 'b sampling show significant differences only for very high discharges. Susp sediment concentration is also dealt with by GURNELL & FENN, but in a p cial environment – a climatic zone about which our knowledge is largely def They find some correspondence between 'englacial' and 'subglacial' flow c nents and the total suspended sediment concentration.

The effects of human interference by ditching in a forest catchment or ment concentration and sediment yield is discussed by BURT, DONOH VANN. A local reservoir afforded an opportunity to monitor in detail the ence of these drainage operations on the sediment concentration which i sed dramatically and, after several months, gradually recovered due to rev tion. TERNAN & MURGATROYD analyse sediment concentrations and fic conductance in a humid, forest and marsh environment. Permanent vege dams are found to influence sediment concentration directly through filt and indirectly through changes in water depth and velocity . Changes in fic conductance are influenced by marsh inputs as well as by the addition o of coniferous forest. The relationship between quality of water and fluvia ment characteristics is dealt with by HERRMANN, THOMAS & HÜBNER analyse the regional pattern of estuarine transport processes. They concluc high pesticide concentrations are correlated with high concentrations of su ded sediment. Hydrodynamic rather than physicochemical factors influen regional distribution in the estuary, and the effect of brooklets draining inten cultivated land is quite evident.

Asher P. Schick

CONTENTS

CATENA SUPPLEMENT 13 p. 109–121 Braunschweig 1988

LANDSLIDE-RELATED FANS IN CALABRIA

M. **Sorriso-Valvo**, Rende

Summary

Fans are common geomorphological features in different environments. Field evidence from Calabria indicates that fans are not always constructed by rivers and recurrent debris flows, but may owe their origin to a single mass-movement. These fans develop through a cyclical four-stage path, which eventually results in an amphitheater-shaped drainage basin with remnants of a fan at its mouth. Reactivation of landsliding may result in the repetition of some stages of the cycle. This paper includes examples illustrating the different phases of fan formation.

Resumen

Los abanicos son formas geomorfológicas frecuentes en los medios de tipo mediterráneo. Por lo que hemos podido estudiar en el sur de Italia es evidente que los abanicos pueden ser no sólo obra de los rios y de los flujos recurrentes de derrubios, sino que pueden también originarse a partir de un único movimiento en masa. La evolución de este último fenómeno tiene lugar a través de un ciclo compuesto de cuatro fases, el cual ocasionalmente tiene como resultado la formación de una cuenca de drenaje en forma de amfiteatro y con restos de un abanico en su desembocadura. La reactivación de los deslizamientos puede dar lugar a la repetición de algunos estadios del ciclo. El artículo incluye algunos ejemplos que ilustran las diferentes fases de la formación de abanicos.

ISSN 0722-0723
ISBN 3-923381-13-1

3-923381-13-1/88/5011851/US$ 2.00 + 0.25

1 Introduction

Alluvial fans are present in regions with diverse climatic conditions. In areas with a Mediterranean climate, these fans are numerous and easily recognizable. They are primarily caused by the sudden drop in transport capacity which streams undergo on emerging from canyons. The various relationships existing between mass wasting on slopes, transport to the stream, carrying power of the stream current, tectonic movements, climate changes, and fan evolution, have been studied over a considerable length of time and it can be said that the mechanisms involved are fairly well known at present (RACHOCKI 1981). It is also known that some fans can owe their construction to recurrent mudflows only (BEATY 1974, KOJAN 1979, RACHOCKI 1981).

Besides the many alluvial fans still in the process of formation, it is quite common in Mediterranean and semi-arid regions to find stream-dissected relict fans. It is probable that these fans were constructed during periods within the Quaternary probably characterized by dryer climatic conditions (HARVEY 1984). Some fans which are still in the construction stage may be found at the end of mud- or debris-flow chutes or at the mouths of basins actively modified by human activities over a long period such as the recurrent forest destruction which has taken place since Classical times in many Mediterranean regions.

Calabria is a typical Mediterranean-climate region where both landsliding and fans are widespread.

Although Calabrian fans are mostly alluvial and a limited number are built by recurrent mudflows or 'sjels' (STARKEL 1976), in certain cases the original starting mechanism was a single landslide of slide-fow type (VARNES 1978).

Some of these have been selected in order to qualitatively illustrate a morphological model to describe how these landslide-related fans evolve. The model depicts a process subdivided into four successive stages:

A - creeping of the mass prone to collapse. This stage can be of long duration.

B - collapse of a rockslide/debris flow type landslide, with the landslide tongue constructing a proto-fan. This stage is very rapid.

C - fan construction by debris flow. The length of this stage depends both on the amount of debris in the scar and the frequency of debris flows.

D - fan dissection when the debris is exhausted. The duration of this stage will obviously depend on the erosion rate and fan dimensions.

The whole sequence may step backwards or forewards, e.g. from stage D to B because of landslide reactivation, or from stage B to D.

2 The Physical Environment of Calabria

The geology of Calabria is characterized by crystalline allochthonous nappes emplaced during the Alpine Orogeny in Miocene time. In northern Calabria, these nappes were overthrust onto Apennine formations. They were subsequently covered by evaporitic and terrigenous sediments during the late Miocene. From the early Pliocene to the Quaternary the basins were filled by thick sediments until tectonic uplifting took place producing normal fault systems which are still active and give the general shape both to the coastlines and mountain ranges. The older the rocks are, the more affected they are by widespread jointing and faulting. Granites and metamorphic rocks are particularly prone to weathering.

Current tectonic activity plays a very important role in intensifying the intrinsic weakness of the parent rocks and the landslide-prone geomorphological conditions: seismic shaking has been directly responsible for the triggering-off of many landslides and the landslide zones generally correspond to those which are highly seismic.

Fig.1 shows that the most of the Calabrian lowlands are characterized by intense rains. Continuous rains, on the other hand, are typical both of the highlands and intramontane valleys (dots to the right at the A+B area). As expected from the morphogenetic approach (FAIRBRIDGE 1968), intense

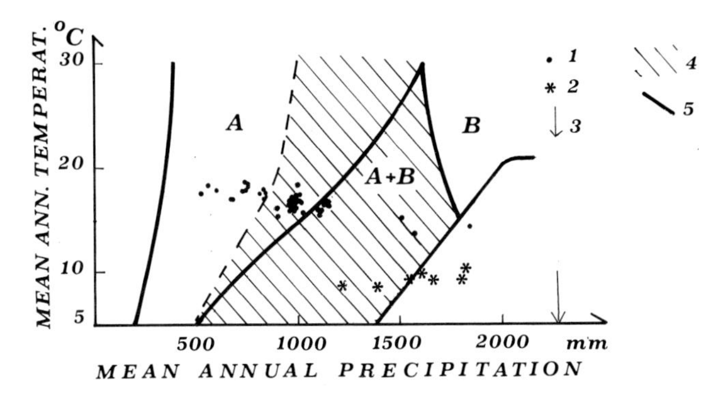

Fig. 1: *Morphoclimatic units characterized by extreme hydrological events and yearly average precipitation- temperature pattern in Calabria.*

Symbols: 1 = raingauges at elevations from 5 to 550 m. 2 = raingauges at elevations from 780 to 1300 m. 3 = maximum average yearly precipitation in Calabria. Morphoclimatic units: 4 = climatic conditions for intensive chemical weathering. 5 = limit for different hydrological extreme events of the types: A = intensive rains. B = continuous rains. Morphoclimatic partioning from STARKEL (1979), simplified.

weathering characterizes most of the region except for the drier areas where the Mediterranean climate is semi-arid. Rapid mass-movement is widespread throughout, with the humid temperature climate in the highlands intensifying soil creep and chemical weathering.

There is a prevalence of erosion and soil slips when intense rains occur, whereas deep landslides are triggered off by continuous rains. However, the present widespread distribution of landslides is the result of previous periods of maximum landsliding so that, while nearly 30% of the Calabrian land surface is affected by landsliding, only 30% of the landslides are still active (CARRARA et al. 1982, SORRISO-VALVO 1985).

The highlands are characterized by forest cover, whereas badlands dominate both the clayey, marly and silty coastal plains and those slopes affected by intensive landsliding in the past.

A large number of fans are to be found at the mouths of canyons where they enter intramontane tectonic valleys or pedemontane and coastal plains. Fans inherited from previous times are dissected. Along the Tyrrhenian coast, some fans could have been tunneled during the first half of this century and have not been affected by the subsequent heavy rains which have damaged large areas of the region. Recent alluvial fans which are still in the constructional phase are also present. After the extreme hydrological event which occured in 1973, the aggradation rate at the apex of one fan being constructed by a large fiumara (FAIRBRIDGE 1968) was about 90 cm per year for the following ten years. This aggradation is due to an increase in the erosion rate caused by both stream erosion and mass-movement.

3 Study Cases

The following cases (fig.2) have been selected from known events involving

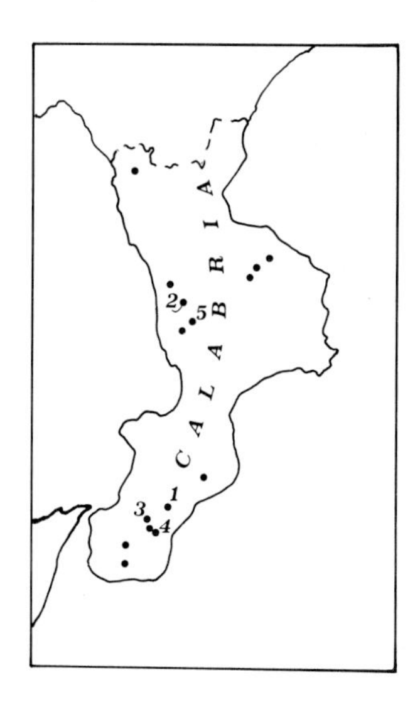

Fig. 2: *Major landslide-related fans in Calabria and study cases.*

1 = Platí. 2 = Lago. 3 = Fiumara Buonamico 1. 4 = Fiumara Buonamico 2. 5 = Aiello.

slopes on different grades of metamorphic rocks. Although this kind of phenomenon can also develop in sedimentary or volcanic rocks, it is seen at its best in metamorphic rocks.

3.1 Fiumara di Platí

The Fiumara di Platí first dissects the high-grade and tectonically crushed metamorphic rocks on the Ionian side of the Aspromonte Range in Southern Calabria and then cuts the Tertiary and Quaternary soft sediments of the pedemontane and coastal lowlands. It is lined with coarse debris and is aggrading along its main stream. Part of its mountainous reach is very asymmetrical (fig.3): the main right-hand slope is a 700 m high fault-scarp while the left-hand one is depressed by a flexure.

The fault scarp displays clear evidence of deep-seated creeping deformation of the "Sackung" type (ZISCHINSKY 1969): joint-related scarps, trenches and sink-holes caused by piping are present in the uppermost part of the slope; crown-shaped scarps, hummocky areas and depressions can be seen in the intermediate slope; a sharp convexity marks the foot of slope where landsliding and sapping are very intensive.

There is little evidence of present movement in the entire deformed mass which is estimated to be in the region of at least 30,000,000 m^3. Topographical monitoring by means of level lines has been recently started in order to detect possible movements and geodetic measurements will begin next year. For the moment, only movements due to superficial landsliding can be detected along the road climbing up the fault-scarp.

3.2 Lago

Near the village of Lago (figs. 2 and 4), a slope carved in Paleozoic phyllites is affected by deep creeping (Sackung). A landslide of slide-flow type caused by creep rupture occurred in unknown historic times. The main body has been almost completely eroded and at present the scarp is retreating by rock toppling, falling and sliding. A fan has been constructed by debris flows at the toe of the main slope (fig.4). The feature covers an area of 0.56 km^2 and is 1,300 m long. According to the inhabitants of the area, mass movement has always been slightly active in the scar although the activity rate is not constant; the most recent accelerations of the mass movement occurred in 1951 and 1984. However, until 1986 very little of the 1984 debris had reached the fan which is being eroded both by the main stream and by the channel draining the bare

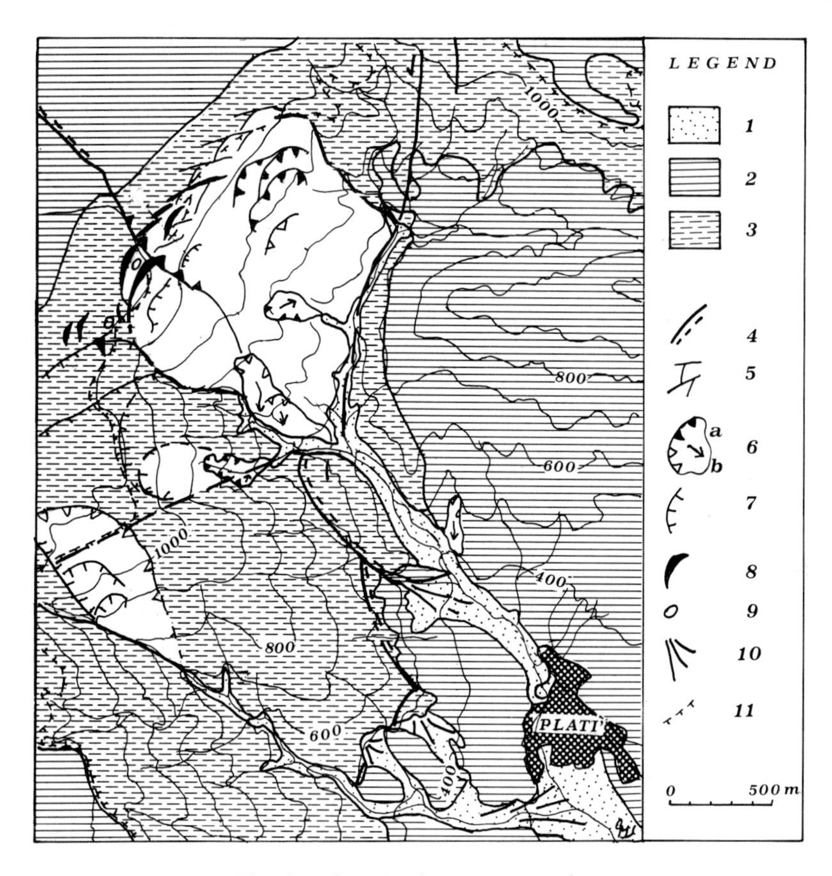

Fig. 3: *The "Sackung" near Platí.*

Symbols: blank = Sackung. 1 = riverbeds. 2 = Tertiary to Quaternary sediments. 3 = metamorphic rocks. 4 = fault (hatchures in the downthrown side). 5 = joint-related scarplets. 6 = main landslide; scarp: a = sharp, b = faint. 7 = minor scarps. 8 = "Sackung" trench. 9 = pseudokarstic sink-hole. 10 = alluvial fan. 11 = edge of marine terrace.

scar. The dissection depth of this channel is about 8 m at the fan apex and gradually decreases to 1 m at the fan tip. There is evidence that a water-rich debris flows overtopped the channel banks and flooded part of the fan. Horizontally-lying tree trunks and 'sieve deposits' made of debris coarser than the underlying fan sediments are to be found midway between the apex and tip of the fan. Overtopping occured once during a six-year observation priod, two years after the landslide reactivation of 1984. Cut-and-fill typical of debris-flows is then in progress. Aggradation is occurring downstream (fig.4). The debris texture can be observed on the walls of the entrenched gully. It is very poorly stratified and blocks of up to 3 m in diameter are mixed with cobble, gravel and sand-sized deposits; this texture is typical of debris-flow material (KOSTASCHUK et al. 1986).

Small remnants of an older fan lie on

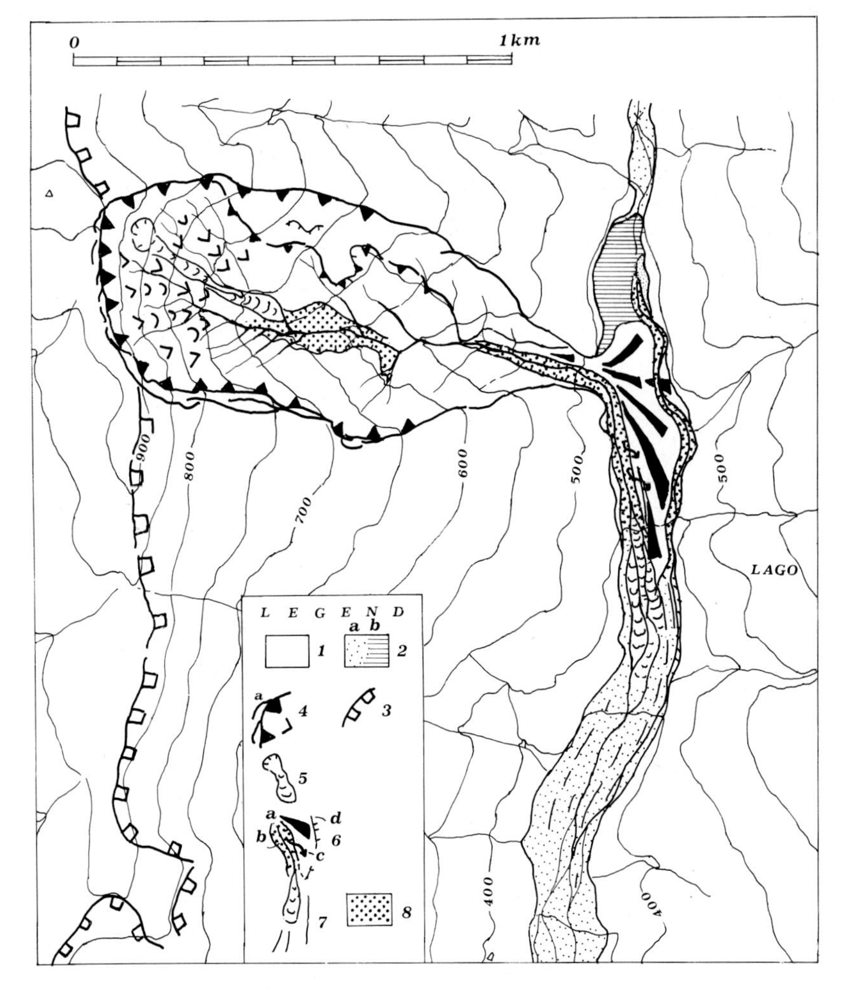

Fig. 4: *"Sackung" and landslide-related fan near Lago.*

Symbols: 1 = low-grade metamorphic rocks. 2 = Holocene sediments; a: riverbeds, b: lacustrine. 3 = rim of the "Sackung". 4 = main scarp; a: tension crack. 5 = debris flow. 6 = landslide-related fan; a: fan, b: entrenched gully, c: debris overflow, d: rim of active erosion zone. z = aggrading riverbeds, 8 = intensive erosion.

the opposite bank, indicating that the present fan is being constructed after a older one had been destroyed. The volume of the present fan is about one fourth of the scar volume. Assuming that the increment of volume of the rock in the loose state could have been compensated by the loss of debris along the mainstream, the older fan must have been up to three times the present one, in volume.

3.3 Fiumara Buonamico

This fiumara dissects the Ionian side of Aspromonte some 10 km SW of the Fi-

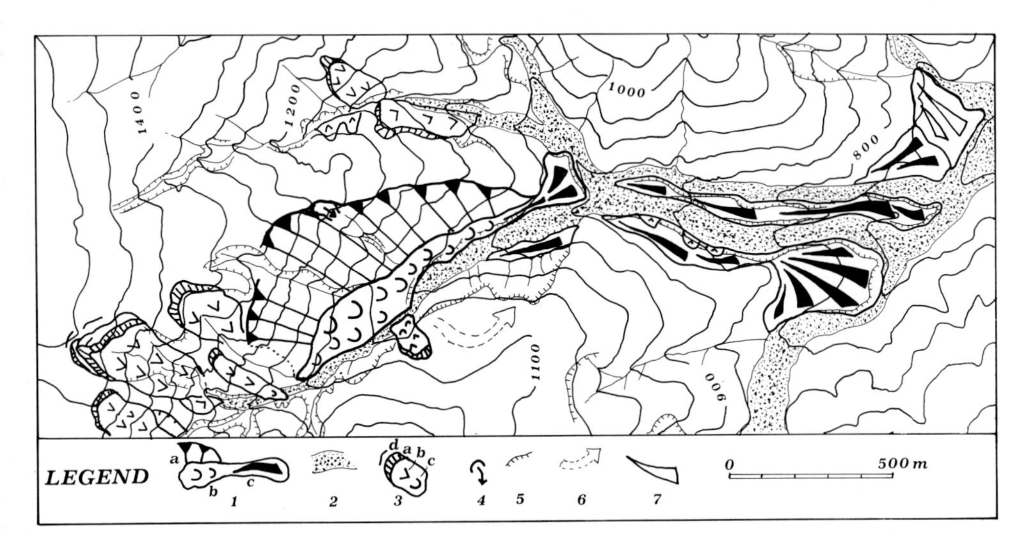

Fig. 5: *Landslide-related fan in the basin of the Fiumara buonamico (Fiumara Buonamico 1).*

Symbols: 1 = landslide of slide/flow type and fan; a: ancient scar, b: debris body, c: fan-shaped tongue. 2 = fiumara riverbed. 3 = recent slide/flow; a: scarp, b: slide, c: flow, d: tension crack. 4 = recent fall. 5 = rim of intensive erosion zone. 6 = hanging valley. 7 = alluvial fan.

umara di Platí. It crosses the same rocks and its slopes are involved in frequent mass movements of the deep creep and landsliding type. Several large landslides occurred during the Holocene as a result of creep rupture (RADBRUCH-HALL 1978, KOJAN & HUTCHINSON 1978) and some in the recent decades, namely in 1951 and 1973.

Fiumara Buoanmico 1 - One of the old phenomena occurred close to the main divide of the fiumara (fig.5) where it mobilized about 9,000,000 m^3 of gneissic rock and constructed a debris tongue channelled along 2,000 m of the main valley. This tongue has a convex transverse profile and ends with a fan-shaped toe characterized by a convex radial profile, since the flowing debris was suddenly stopped as it reached a wider section of the fiumara bed. Fig.5 shows a secondary fan where a left-hand tributary flows into the fiumara. The main fan is c350 m long and has an average gradient of c20%; the secondary fan is c250 m long and has a gradient of c30%; the debris tongue has a gradient if 14%. The terrace height is rather regular and ranges between 20 and 30 m. Erosion subsequent to landslide activity has entrenched the fiumara into the debris tongue and main fan, resulting in a terraced fan.

Fiumara Buonamico 2 - About 8 km downstream of the site described above, a rockslide/debris flow was triggered off by an extreme continuous rain in 1951 (fig.6). The rock mass displaced was c7,000,000 m^3 over an area of 0.35 km, i.e., with

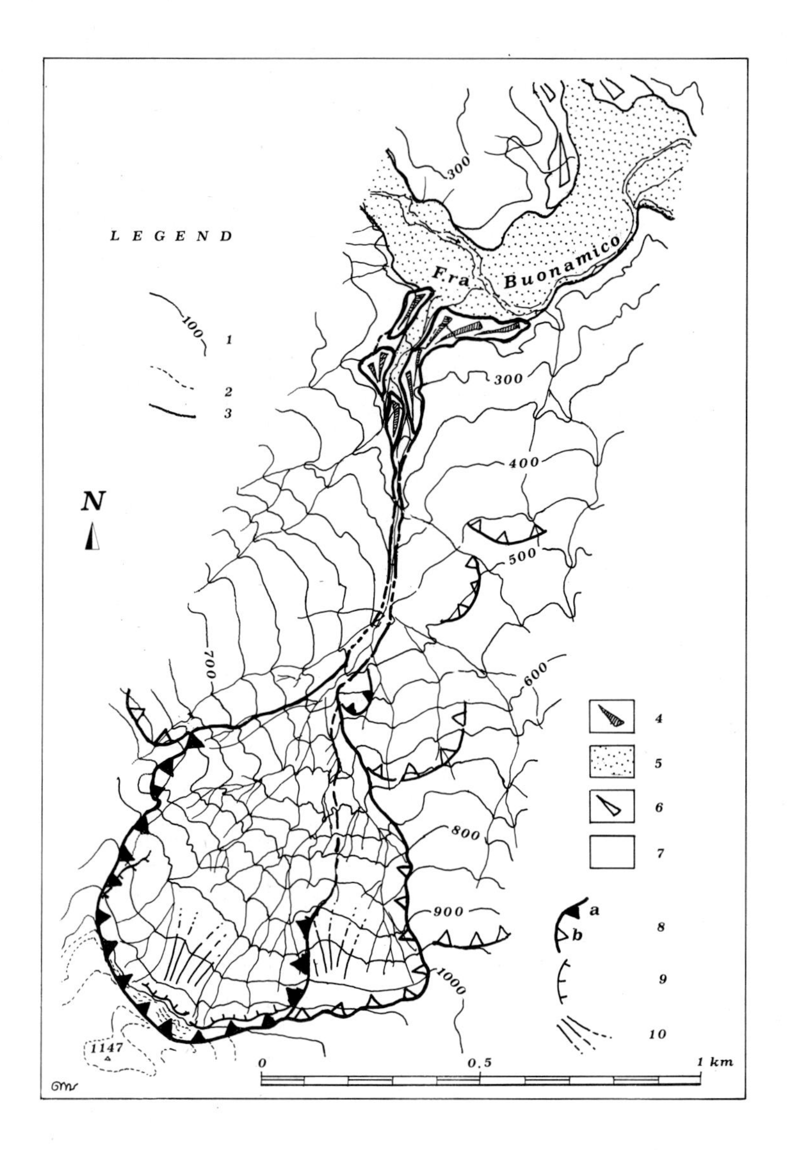

Fig. 6: *Amphitheatre landslide in the basin of Fiumara Buonamico (fiumara Buonamico 2).*

Symbols: 1 = contour line. 2 = approximative contour line. 3 = landslide limit. 4 = landslide-related fan. 5 = riverbeds. 6 = alluvial fan. 7 = high-grade metamorphic rocks. 8 = main scarp; a: sharp, b: smoothed. 9 = secondary scarp. 10 = talus.

an average thickness of only 20 m. this has been confirmed by comparing topographic maps obtained from aerial photographs taken in 1941 and 1978. Since the maximum width of the crown is 470 m, the width/thickness ratio is .0042. This extremely low value is unexpected for rockslides and indicates that collapse probably propagated from a smaller landslide throughout the present scar that now appears as a bare, amphitheatre-shaped basin leading through a steep and narrow canyon to the Buonamico riverbed. Remnants of a small debris fan are present in the confluence area (fig.6); the total volume of the fan is c100,000 m^3. In this case, most of the landslide debris was cleared off ba the flood in the main stream so that only a small fan could form. This event was in fact a reactivation of an old, stabilized landslide as is the general rule all over Calabria where only a few of the landslides are primary events (CARRARA et al. 1982).

3.4 Aiello Calabro

In Northern Calabria, debris flows from a huge and old (Holocene) landslide have constructed a large fan (fig.7). The landslide area is more than 1.9 km^2 and the length of the scar-fan complex is more than 4 km. This feature is a clear example of a fan which was rapidly constructed simultaneously with and subsequently to a large mass-movement. The main stream must have been dammed, as is evidenced by the presence of lacustrine deposits upstream of the fan. Its shape is regular in its upper portion while the distal part is digitated where it adapted to morphology of the valley-bottom. The average gradient is 10%. Large lobes of 'sieve' deposits formed very steep (30%) secondary fans which filled some small side valleys after overriding the lowermost ridges. This fan complex is at present dissected and its surface is relict. The debris, which is very poorly statified, has the typical texture of debris-flow deposits. The core of the fan was probably formed by the tongue of the large landslide followed by a seris of debris flows which created the fan surface. The tongue of the main landslide body crops out from beneath the apex of the fan; its remnants, which are up to 50 m thick, also occur in the scar of the landslide.

4 Discussion

As already mentioned in the Intrudoction, the cases described can be ordered into a sequence which corresponds to an evolutionary path. This path, which is schematically shown in fig.8, can be divided into four main stages:

A — Creep

The first stage is illustrated by the case found near Platí. The dominant process on the slope is rock flow (Sackung). Superficial landslides can develop mostly in the lower portion of the slope wheras a set of other diagnostic superficial deformations can extend throughout (NEMCOK 1972, RADBRUCH-HALL 1978, SORRISO-VALVO 1979). The duration of this phase can be very long; indeed, some of the phenomena observed appear to have been in this condition since early Holocene at least. Creep can stop temporarily or forever even if it is extremely sensitive to long-term climatic change,

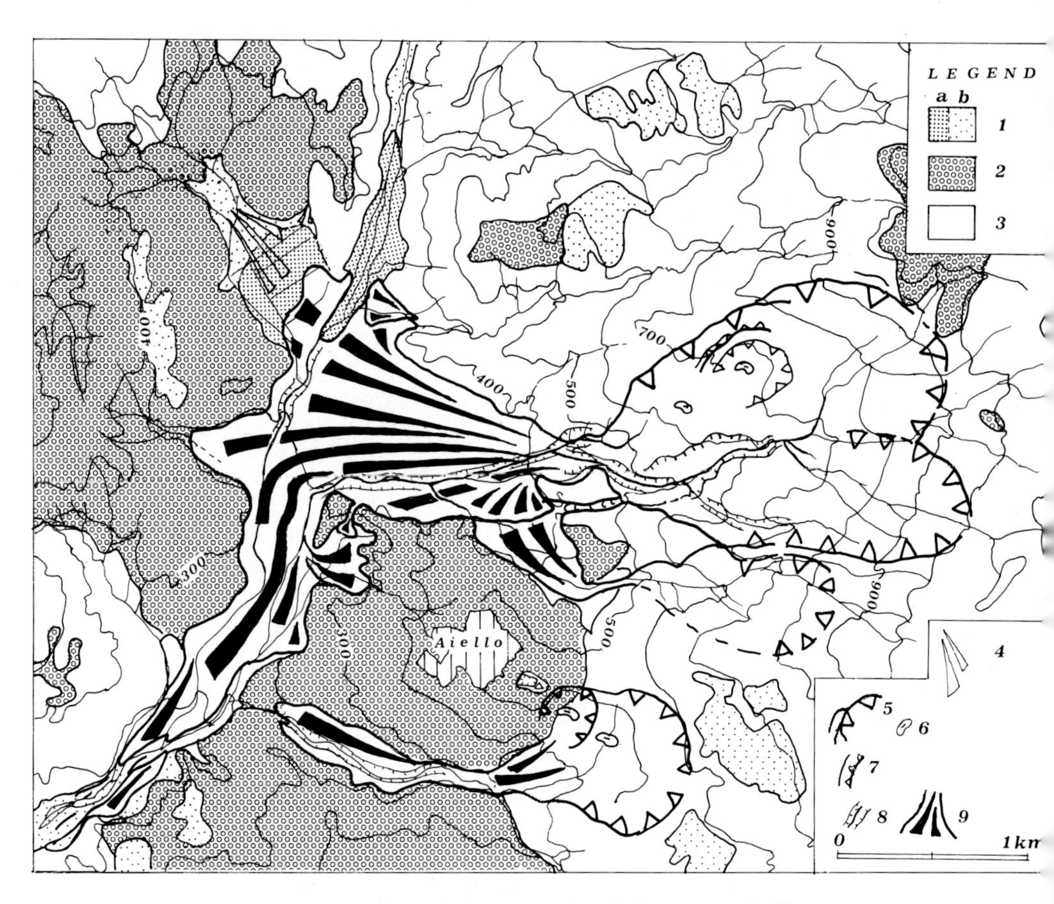

Fig. 7: *Ancient landslide-related fans near Aiello.*

Symbols: 1 = Quaternary sediments; a: lacustrine, b: riverbeds and terrces. 2 = Miocene sediments. 3 = low-grade metamorphic rocks. 4 = alluvial fan. 5 = main scarp and tension cracks. 6 = undrained depression. 7 = uphill facing scarp. 8 = entrenched gully. 9 = landslide-related fan.

to extreme hydrological events, and to tectonic activity. Due to the possible interplays between these factors, it is impossible to predict when a subsequent phase might begin.

B — Collapse

Rockslide/debris flow suddenly occurs when creeping progresses from the secondary to tertiary stage (EMERY 1978). The collapse may result in an amphitheatre-shaped scar and in a landslide body which takes the shape of a fan as in the cases of Fiumara Buonamico 2 and Lago (figs.4 and 6). Sometimes, the landslide body can be simultaneously or subsequently removed by intensive erosion, as in the Fiumara Buonamico 2 and 1 cases, respectively (figs.6 and 5), or else it can be channelled along the main stream as will probably happen in the case in the Fiumara di Platí (fig.3).

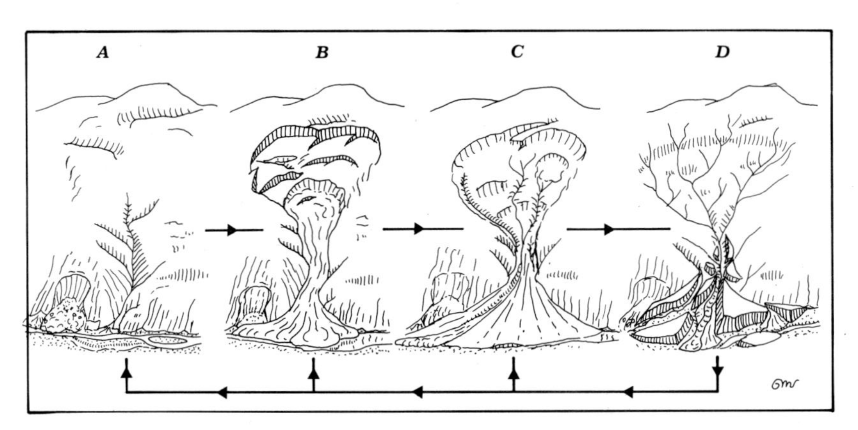

Fig. 8: *Evolution cycle of an hypothetical landslide-related fan.*

Arrows indicate the possible steps. the shape, though indicative, is typical of parent rocks that hebave as ductile at the same scale as that of the main slope. In brittle and jointed rocks the main flow of stage B may take the form of a high-speed debris avlanche (fig.3). The shown slope morphology prior to the onset of the mass-movement, is purely indicative.

C — Fan construction

Both the weakened parent rock and the debris remaining in the scar are mobilized by mass movement subsequent to a reactivation of sliding and/or by debris-flows. These processes can build a fan in a relatively short term.

The Lago case (fig.4) illustrates present fan construction by means of debris flow. The final dimensions of the fan depend not only upon the amount of material available and on the transport capacity of debris-flows, but also on the erosional power both of the main stream and that dissecting the fan itself. The possible range is well illustrated by Buonamico 2 and Aiello (figs.6 and 7).

D — Fan destruction

As the slid material is exhausted by erosion and debris flow, streams flowing from the scar attain a higher erosional power than in the earlier stages and this power is expended in eroding the fan. Erosion starts from the apex and expands towards the tip. The fan surface may be attacked by other gullies and by the main valley stream. The scar may be dissected by surface erosion caused by a developing network of rills and gullies. The Lago and Fiumara Buonamico 2 scars are good examples of such drainage networks. During this final phase the site once affected by a landslide appears as an amphitheatre-shaped basin with a dendritic drainage network and remnants of a fan at the confluence with the main stream. Lacustrine deposits may occur along the main valley, upstream of the fan, as the result of damming.

The sequence described may proceed along a path that steps backwards and partially repeats itself, when reactivation of mass-movement occurs.

A path of type A—B—C—D—B—C—D... represents the most common evolution of the process. In this case, the volume of the fan can be very much smaller than that of the rock removed from the scar. The case of Lago is a good illustration of such a path.

The dynamics of this cycle depend both upon the amount and seasonal distribution of the precipitation (RACHOKI 1981) and upon the quality of the parent rock. In alluvial fans, the dissection of their surface is an important indicator of climatic (RACHOKI 1981), tectonic and other environmental changes (HARVEY 1984) while in debris-flow fans dissection can be caused by the flows themselves (BEATY 1974, KOSTASCHUK et al. 1986). In landslide-related fans the channel flowing from the scar can become entrenched for different fans in the same area or at different times on the same fan. This indicates either that the fan cycle is in stage C (fig.4) where debris flow predominates, or that the cycle is in stage D (fig.6). The extreme hydrological events characterizing Mediterranean climates can account for the recurrent reactivation of mass movements which sets the fan cycle back from stage D to stage B or C. The frequency of such events in Calabria is roughly five times per century (ERGENZINGER et al. 1978). Obviously not all potential landslides are activated at each event. Consequently, landslide-related fans are to be found at different stages of their evolutionary cycle.

The importance of rock quality is not to be underestimated since rock flow (Sackung) and rockslide/debris flows appear to be more frequent and well developed on low grade metamorphic rocks and/or densely jointed rocks (SORRISO-VALVO 1979, 1984, 1985).

As expected, these landslide-related fans also occur elsewhere. An example of a case in stage C has been reported by KOJAN (1979) at Wrightwood, California. There, lowgrade metamorphic rocks are involved and a holiday-estate village is threatened by recurrent debris flows up to 2,000 m long. VERSTAPPEN (1983, 209) reports another stage C case in the Himalayan foothills in India. In Northern Algeria, at Oued-Cheliff, a 300 m long earth slide/flow involving Miocene siltstones has a fan-shaped tongue which has been little affected by subsequent erosion. It is in a slowly progressing stage without experiencing stage C. The Valle del Bove on Mount Etna thought to be a relic caldera, is probably a huge complex of Holocene landslide scallops with a fan of at least 2 km lying at its mouth (GUEST et al. 1984).

Other phenomena, such as natural dam failure (COSTA 1983), or major storms, may be responsible for fan generation. These fans may also appear as isolated features, as it may happen for landslide-related fans. In these cases, however, evidence of single-event genetation should be found. The possibility of morphological convergence between single-event and landslide-related fans is high. This point undoubtly needs further study.

Acknowledgement

I thank Adrian Harvey and one anonymous referee for generous and valuable advice.

References

BEATY, C.B. (1974): Debris flows, alluvial fans and revitalized catastrophism. Z. Geomorph. N.F., Suppl. **21**, 39–51.

CARRARA, A., SORRISO-VALVO, M. & REALI, C. (1982): Analysis of landslide form and incidence by statistical techniques, Southern Italy. CATENA, **9**, 35–62.

COSTA, J.E. (1983): Paleohydraulic reconstruction of flash-flood peaks from boulder deposits in the Colorado Front Range. Geol. Soc. America Bull., **94, 8**, 986–1004.

EMERY, J.J. (1978): Simulation of slope creep. In: VOIGHT, B. (Ed.), Rockslides and Avalanches, **1**. Elsevier, Amsterdam.

ERGENZINGER, P., GORLER, K., IBBEKEN, H., OBENAUF & P., RUMOHR, J. (1978): Calabrian Arc and Ionian Sea: Vertical Movements, Erosional and Sedimentary Balance. In: H. CLOSS, D. ROEDER, K. SCHMIDT (Eds.), Alps, Apennines, Hellenides. E. Schweizerbart'sche Verlagsbuchhandlung, Stuttgart.

FAIRBRIDGE, R. (1968): the Encyclopedia of Geomorphology. Reinhold Book Corp., N.Y.

GUEST, J.E., CHESTER, D.K. & DUNCAN, A.M. (1984): The Valle del Bove, Mount Etna: its origin and relation to the stratigraphy and structure of the volcano. Journ. Volcanology and Geothermal Research, **21**, 1–23.

HARVEY, A.M. (1984): Aggaradation and dissection sequences on Spanish alluvial fans: influence on morphological development. CATENA, **11**, 289–304.

KOJAN, E. (1979): The mudflow-landslide hazard at Wrightwood, San Bernardino County, California. Forest Service, U.S. Depart. of Agriculture, Pleasant Hill.

KOJAN, E. & HUTCHINSON, J.N. (1978): Mayunmarca rockslide and debris flow. Peru. In: VOIGHT, B. (Ed.), Rockslides and Avalanches, **1**, Elsevier, Amsterdam.

KOSTASCHUK, R.A., MACDONALD, G.M. & PUTNAM, P.E. (1986): Depositional process and alluvial fan-drainage basin morphometric relationships near Banff, Alberta, Canada. Earth Surface Processes and Landforms, **11**, 471–484.

NEMCOK, A. (1972): Gravitational slope deformation in high mountains. Proc., Journ. Soil Mechanics Found. Div., **93**, SM4, 403–417.

RACHOCKI, A.H. (1981): Alluvial fans. J. Wiley & Sons, Chichester.

RADBRUCH-HALL, D. (1978): Gravitational creep of rock masses on slopes. In: B. VOIGHT (Ed.), Rockslides and Avalanches, **1**, 607–675, Eslevier, Amsterdam.

SORRISO-VALVO, M. (1979): Trench features on steep-sided ridges of Asptromonte, Calabria (Italy). Proceed. Polish-Italian Seminar on Superf. Mass-movement. Szymbark, Ist. Godsp. Wodnej, 99–108.

SORRISO-VALVO, M. (1984): Deep-seated gravitational slope deformations in Calabria (Italy). Actes Coll. Mouvem. Terrain, Docum. B.R.G.M., **83**, 81–90.

SORRISO-VALVO, M. (1985): Mass movement and slope evolution in Calabria. Proceed. IV I.C.F.L., 23–30, Tokyo.

STARKEL, L. (1976): The role of extreme (catastrophic) meteorological events in contemporary evolution of slopes. In: DERBYSHIRE, E. (Ed.), Geomorphology and Climate. J. Wiley & Sons, Chichester, N.Y., Brisbane, Toronto.

VARNES, D.J. (1978): Slope movement types and processes. In: SCHUSTER, R.L. & KRIZEK, R.T. (Eds.), Landslides Analysis and Control. National Academy of Sciences, Washington, D.C.

VERSTAPPEN, H.T. (1983): Applied Geomorphology. Elsevier, Amsterdam.

ZISCHINZKY, U. (1969): Über Sackungen. Rock Mechanics, **1**, 30–52.

Address of author:
Marino Sorriso-Valvo
CNR-IRPI
via G. Verdi, 1
87030 Roges di Rende
Italy

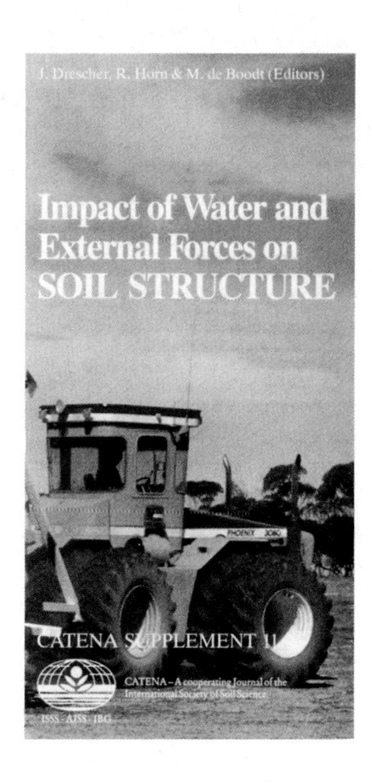

J. Drescher, R. Horn & M. de Boodt (Editors):

Impact of Water and External Forces on SOIL STRUCTURE

CATENA SUPPLEMENT 11, 1988

DM 149,– / US $88.–

ISSN 0722-0723 / ISBN 3-923381-11-5

CONTENTS

CATENA SUPPLEMENT 13 p. 123–137 Braunschweig 1988

CONTROLS OF ALLUVIAL FAN DEVELOPMENT: THE ALLUVIAL FANS OF THE SIERRA DE CARRASCOY, MURCIA, SPAIN

A.M. **Harvey**, Liverpool

Summary

The influence of three set of factors, tectonic, climatic and dynamic on the geomorphology of the alluvial fans of the Sierra de Carrascoy, Murcia, is reviewed in the wider context of southeast Spain. Tectonic factors have a long term influence on the development and location of alluvial fans but tectonic disturbance has only a limited and local influence over medium and shorter timescales on fan devlopment and morphometry. Climatic factors influence the sediment production from mountain source-areas and hence the medium term sequences of alluvial fan aggradation and dissection, within the context of longer term progressive changes in fan morphology. Dynamic factors control the local short-term relationship between sediment supply, erosion and deposition, and fan morphometry. Mid-fan headcut development and distal trenching, resulting from particular combinations of process/form relationships may ultimately lead to the complete dissection of the fan surface.

ISSN 0722-0723
ISBN 3-923381-13-1

D–3302 Cremlingen-Destedt, W. Germany
3-923381-13-1/88/5011851/US$ 2.00 + 0.25

Resumen

El artículo presenta una revisión de la influencia de tres conjuntos de factores, tectónicos, climáticos y dinámicos, en la geomorfología de los abanicos aluviales de la Sierra de Carrascoy, Murcia, situándolos en el contexto más amplio del sudeste de España. Los factores tectónicos tienen una influencia a largo plazo en el desarrollo y localización de los abanicos aluviales; esta acción tectónica condiciona sin embargo sólo de manera limitada y local, a escala media y corta, el desarrollo y morfometría de los abanicos. Los factores climáticos influyen en la producción de sedimento en la cuenca, y por tanto en las secuancias a medio plazo de la agradación y disección de los abanicos aluviales dentro del contexto, a más largo plazo, de los cambios progresivos de la morfología de los abanicos. Los factores dinámicos son los que controlan las relaciones locales, a corto plazo, entre suministro de sedimento, erosión, depositación y morfometría de los abanicos. El desarrollo de procesos de erosión, que se incian en la cabecera y en la parte distal del abanico, resultantes de combinaciones específicas de las relaciones forma/proceso, pueden llegar a producir una total disección de la superficie del abanico.

1 Introduction

Alluvial fans play an important role in the geomorphology of dry-region fluvial systems, their aggradational or dissectional behaviour influencing the strength of coupling and continuity between mountain sediment-source areas and basinal main drainages. Aggrading fans may act as buffers within the fluvial system trapping coarse sediment and reducing the linkage through the system (BRUNSDEN & THORNES 1979), whilst during dissection, especially the total-dissection of fan surfaces, there may be continuity of sediment movement from source-area to main drainage. Three sets of factors may be identified, influencing the aggradation/dissection relationships during fan development. These are:

i) tectonic factors, influencing the overall relief of the system and the rate at which base-level related dissection proceeds;

ii) climatic factors, influencing the sequence of sediment availability, and

iii) dynamic factors, influencing the within-fan relationships between sediment supply, fan morphology and erosion and deposition.

This paper examines the influence of these three sets of factors on the geomorphology of the Quaternary alluvial fans of the Sierra de Carrascoy, Murcia (fig.1), within the general context of fan development in southeast Spain as a whole.

2 The Sierra de Carrascoy

The Sierra de Carrascoy forms part of the internal zone of the Betic Cordillera (RIOS 1978, ALVARADO 1980), and consists of Permo-Triassic low-grade metamorphic, sedimentary and igneous rocks, folded into comlex nappe structures (BODENHAUSEN & SIMON 1965). Emplacement of the major Betic nappes pre-dates the middle Miocene, by which time sedimentation was taking place in basins between uplifted mountain ranges. The Neogene depositional sequence was dominantly marine in the late Miocene and early Pliocene followed by terrestrial sedimentation in the late Pliocene and early Pleistocene. Continuing Neogene tectonic activity is evidenced by the incorporation of upper Miocene rocks into the Carrascoy structures, and continued uplift of the block relative to the surrounding basins is revealed by the differential deformation of upper Miocene, Pliocene and early Quaternary rocks, decreasing in intensity up the sequence and southwards away from the Sierra de Carrascoy into the Cartagena lowland (I.G.M.E.).

Following the Pliocene uplift and emergence, Quaternary tectonism has been dominated by movement along major strike-slip faults (BOUSQUET 1977), one of which bounds the north side of the Guadalentin valley (fig.1, inset). Within the Carrascoy block compressional tectonics have created high-angle reverse faults separating the Carrascoy mountain-front from the southern margin of the Gualadentin valley.

During the Quaternary, alluvial fans accumulated in three zones on the margins of the Sierra de Carrascoy; at the faulted northern mountain-front (fig.1, Group 1), in a structurally controlled inter-montane depression south of the Carrascoy crestline (fig.1, Group 2), and mantling pediment surfaces on the southern and western margins of the

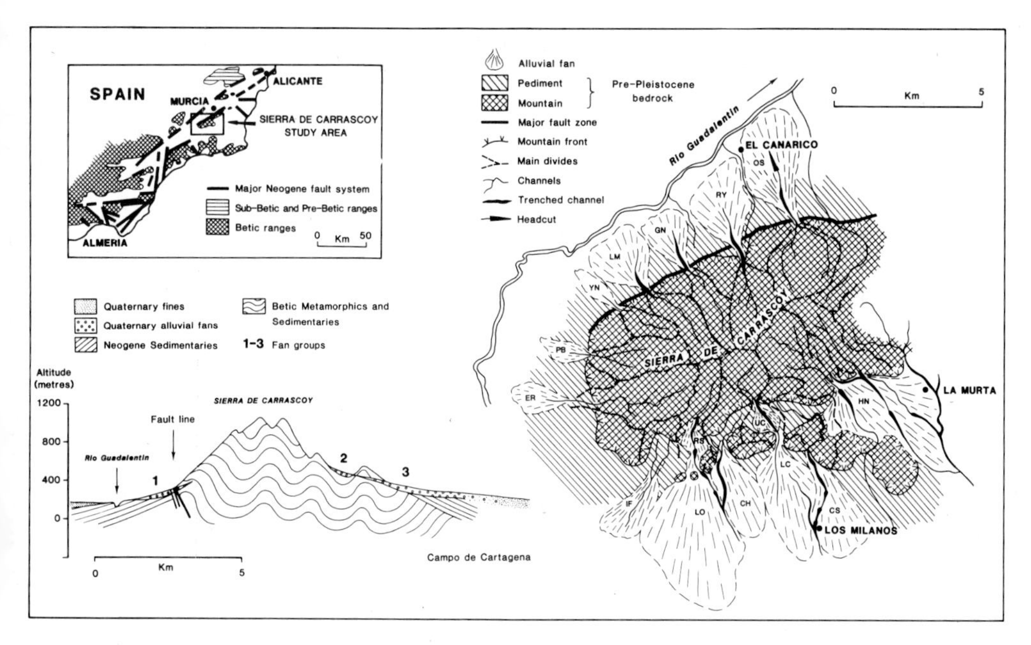

Fig. 1: *Alluvial fans of the Sierra de Carrascoy: location map and north-south cross section.*

1: Northern mountain-front fans (OS, Oscuro; RY, Roy; GN, Ginesa; LM, La Murta; YN, Yncholete.)
2: Inter-montane fans (HN, Heradon; UC, Upper Carrascoy; RS, Ros.)
3: Southern and Western pediment-mantling fans (CS, Corachos; LC, Lower Carrascoy; Ch, Charco; LO, Loberos; IF, Infirno; ER, Romero; PB, Penas Blancas.)
Names are from 1:50,000 Mapa Militar de Espana. Insert shows location and tectonic setting within southeast Spain, Fault lines after BOUSQUET (1979).

mountains (fig.1, Group 3). These latter form the proximal parts of the extensive 'glacis' forms of the Campo de Cartagena (DUMAS 1977), the alluvial fan deposits radiating away from the mountain catchments at the fan apices and burying the pediment surfaces by depths in excess of 20 m (HARVEY 1984a).

3 Factors Influencing Alluvial Fan Development

3.1 Tectonic Factors

Tectonic uplift has been identified as a major factor in the development of alluvial fans (BULL 1964, 1977) both in the context of the creation of the original relief and in influencing fan sedimentation. In southeast Spain the post-Pliocene uplift of the major mountain ranges relative to the intervening basins has created a topography suitable for the accumulation of mountain-front fans. This is clearly the case along the northern margin of the Sierra de Carrascoy (fig.1, photo 1). To the south however, there is no obvious mountain-front fault line and the fans rest on pediment surfaces cut across pre-Pleistocene rocks, which since the early Quaternary have been gently tilted towards the Campo de Cartagena (BIROT

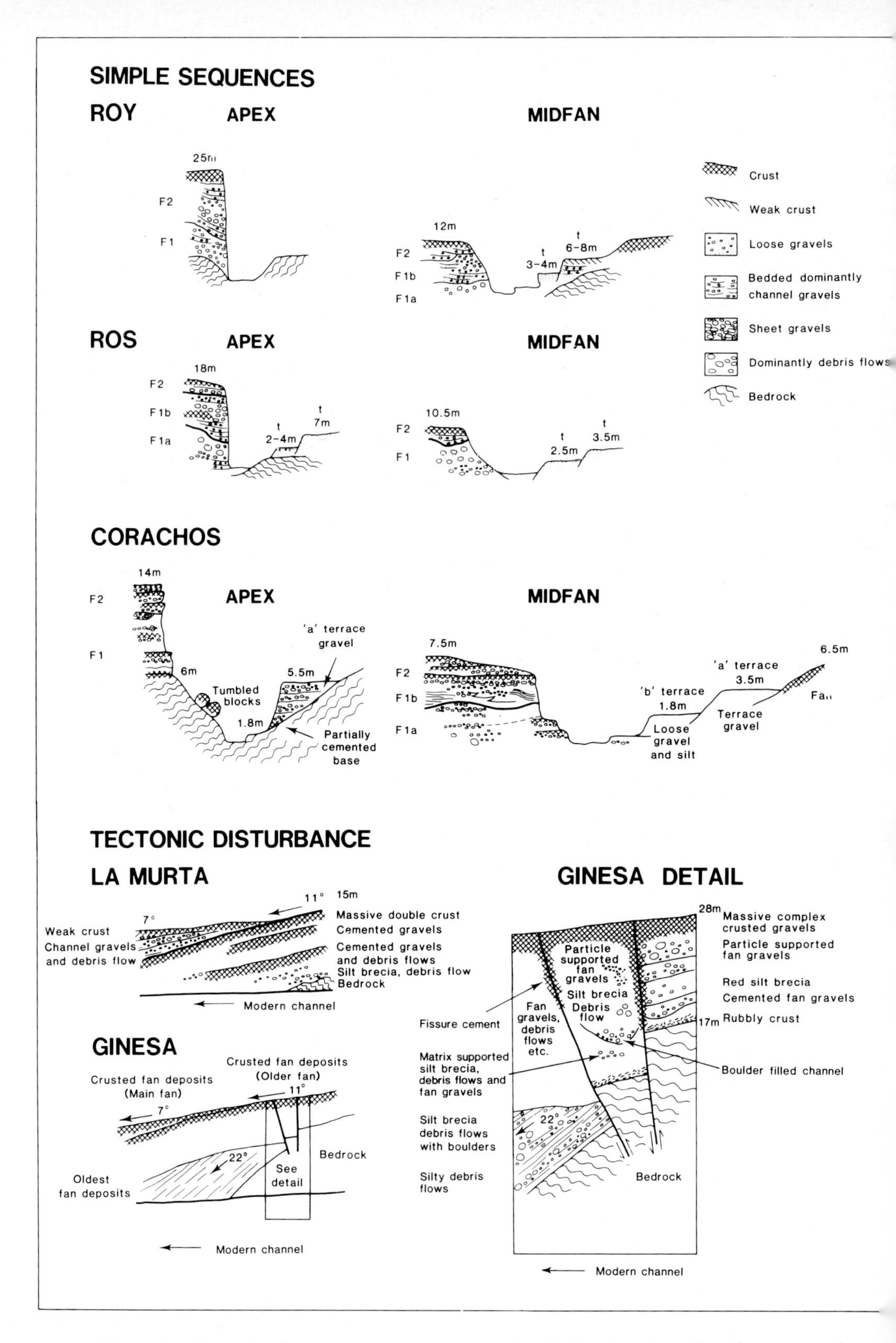

Fig. 2: *Example fan sequences at apex and mid-fan locations for three example fans from Group 1 (Roy), Group 2 (Ros) and Group 3 (Corachos). Below is shown evidence of tectonic disturbance on Group 1 fans.*

Photo 1: *Roy fan from above. Note (1) fanhead trench cut c20 m below fan surface, (2) intersection point (3) distributary channel system on distal fan surface.*

& SOLE SABARIS 1959).

One result of regional uplift may be the rapid incision of major drainages, which may cause base-level dissection of mountain-front fans. Further south, in Almeria province such dissection has prevented the accumulation of alluvial fans near main drainages (HARVEY, in press). In the Carrascoy region the limited incision of the Rio Guadalentin has not caused such dissection of the distal portions of the Carrascoy mountain-front fans and there is no equivalent major drainage to the south, so no base-level induced dissection is present there either.

Tectonic activity during and after fan sedimentation may be evidenced by disturbed and faulted fan deposits. On two of the Carrascoy mountain-front fans such disturbance is evidenced in sections exposed by incision of fan-head trenches (fig.2). Near the apex of La Murta fan, older well-cemented fan deposits dip away from the mountain front at c11° and are overlain by younger deposits lying at c7°, forming the aggradational surface of the fan. At Ginesa three sets of deposits are visible dipping at c22°, c11° and c7°, the youngest again forming the fan surface. The middle group, forming the older fan surface, have been demonstrably disturbed by two high angle reverse faults (fig.2). The deposits are capped by a massive well-indurated calcrete crust from which $CaCO_3$ cement penetrates fissures along the fault planes. There has apparently been no movement along these faults since crust formation and cementation. Nearby, a similar fault is exposed in a roadcut, where Triassic soft gypsiferous marls have diapirically risen along the fault plane and disturbed the Quaternary fan deposits.

Such fault disturbance is not always evident. On the neighbouring Roy fan no disturbed sections could be found along the walls of the fanhead trench. South of the mountains there is no evidence of tectonic disturbance in fanhead trench sections in either the intermontane or the southern group of fans.

3.2 Quaternary Sequences

As elsewhere in southeast Spain the Carrascoy fans show evidence of early phases dominantly of aggradation, culminating in the formation of calcrete-crusted fan surfaces, followed by later phases dominantly of dissection (HARVEY 1984a). During both phases aggradation and dissection were episodic, aggradational tendences reflecting excess sediment generation and dissectional tendencies, periods of low sediment generation.

Deposits of aggradational phases show thicknesses of up to and over 20 m exposed in fanhead-trench sections, often resting on erosional pediment surfaces cut across pre-Pleistocene rocks Facies present are characteristic of dry-region alluvial fans (BLISSENBACH 1954, HOOKE 1967), and range from debris-flow to fluvial deposits (HARVEY 1984b). True matrix-supported debris-flow deposits are common in the smaller fans especially in proximal locations. Almost structureless clast-supported gravels occur in both proximal and distal locations and appear to be related to two depositional mechanisms; sheetflood deposition, especially in distal environments, and in proximal environments, deposition by wet debris flows, transitional or hyperconcentrated flows from which the matrix drained on deposition (WELLS & HARVEY 1987). In distal environments, especially in the larger fans of the southern Carrascoy group, silt sheets occur interbedded with thin gravel sheets, again suggestive of sheet flooding. True fluvial channel gravels are present in both proximal and distal environments, showing obvious characteristics of channel deposition, such as stratification, imbrication and channel-fill structures.

The sections exposed in fanhead and through-fan trenches show sedimentary sequencies of at least two major episodes of aggradation, separated by buried calcrete crusts, apparently representing surface stability and soil formation (BUTZER 1964, ROHDENBURG & SABELBERG 1980), or by erosional horizons representing dissectional episodes. The earlier phases (F_1, on the examples shown on fig.2) are dominated by debris-flows proximally and silt and gravel sheets distally. The later phases (F_2 on the examples shown in fig.2) are dominated by channel gravels. This type of sequence, which is also characteristic of other areas of southeast Spain (HARVEY 1978, 1984a), suggests a progressive diminution in the availability of fine sediments from the mountain source-areas. Since the aggradation phases, calcrete crusts have formed, capping the sequences.

During the later phases of fan development dissection by fanhead and through-fan trenches has been dominant, but intermittent, with at least two major depositional episodes represented by terrace deposits within the fanhead trenches and which in part spread onto the dis-

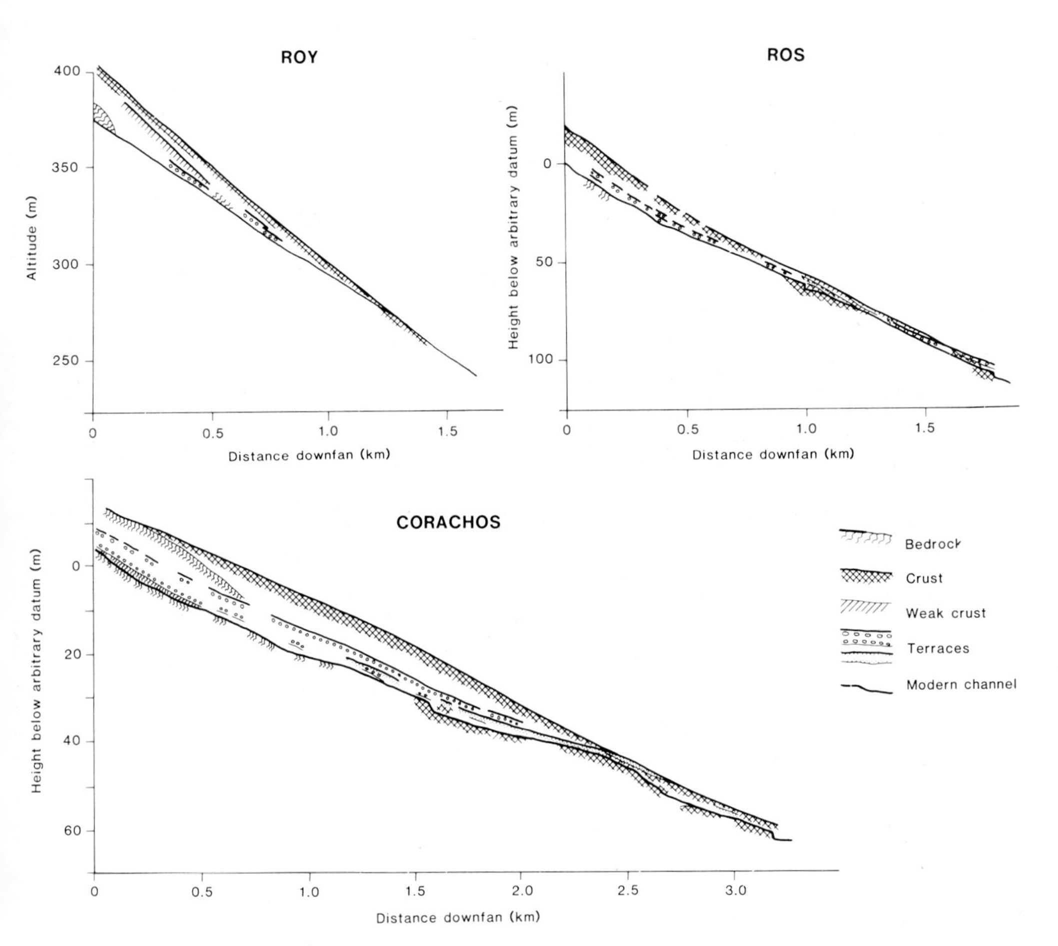

Fig. 3: *Fan profiles, three example fans. Note scale differences. Roy fan has a profile of type C, Ros and Corachos of type E (see fig.4).*

tal fan surfaces (fig.2, 3, photo 2). The terrace deposits are dominantly of fluvial channel gravels. The earlier gravels (especially terrace 'a' on Corachos, fig.2) show varying degrees of cementation at the bedrock contact and locally weak or immature calcrete crusts at the surface. The younger gravels, which locally include interbedded silts are loose and unconsolidated.

No major differences in the fan sequences, during either aggradation or dissection phases can be identified between the northern and the two other fan groups. As a whole the fan sequences appear to relate to a considerable period during the Quaternary. The fans are inset below and therefore post-date early Quaternary sediments in the Sucina area of the Campo de Cartagena (I.G.M.E.) Mature calcrete crusts of the type characteristic of the upper fan surfaces show complex profiles, with massive indurated petrocalcic horizons often showing evidence of brecciation and secondary recementation, which according to the cri-

Photo 2: *Corachos fan, fanhead trench looking downfan, with terraces (a and b) inset below fan surface (f).*

teria used by DUMAS (1969), suggest at least a Riss-age. Crust development on the oldest terrace surfaces of the dissection phases is much less mature, corresponding to that identified by DUMAS (1969) elsewhere in southeast Spain as Wurm-age. This would suggest that the episodic aggradational phases relate to a period up to the middle Pleistocene and the episodic dissectional phases relate to the Würm and the Holocene. This accords with the sequence suggested for the alluial fans in the Pre-Betic region of Alicante (HARVEY 1978). Further south towards Almeria, where the dissection sequence is simpler and crust development on the fan surfaces is less, the aggradation phases may have continued into the Würm (HARVEY, in press).

The modern climate is semi-arid with annual precipitation totals of c300 mm (GEIGER 1970) and a marked seasonality, with most precipitation falling in autumn and spring. Quaternary climates appear to have been drier, with insufficient moisture during 'glacials' for tree growth (AMOR & FLORSCHUTZ 1964), but presumably also had a marked seasonality. This seasonality is important both from the point of view of sediment production and of calcrete formation (BUTZER 1964). Throughout the western Mediterranean major periods of sediment generation during the Pleistocene appear to relate to cold, dry 'glacials' (VITA FINZI 1972, CUENCA PAYA & WALKER 1976, ROHDENBURG & SABELBERG 1980, SABELBERG

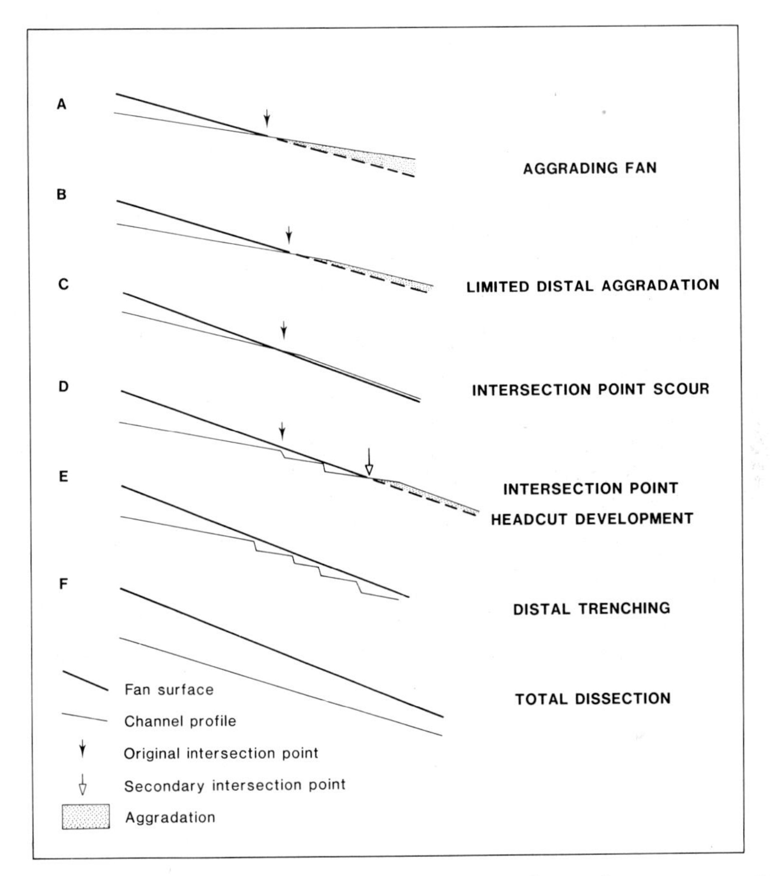

Fig. 4: *Schematic model for fan and channel profile relationships on aggrading and dissecting fans (after HARVEY 1987).*

1977). Within long-term fan aggradation and dissection phases, shorter-term aggradational episodes probably relate to excess sediment production during dry 'glacials', perhaps as the result of the greater effectiveness of extreme seasonal rains under more arid conditions (BAKER 1977). An exception to this is the youngest Holocene terrace, which relates to human-induced erosion (VITA FINZI 1972). Since the late Pliocene to early Quaternary uplift, the progression from dominant aggradation to dominant dissection together with the progressive diminution in the availability of fine sediment suggest a long-term trend of progressive exhaustion of sediment supply from the mountain source areas (as first suggested by ECKIS 1928); onto which are superimposed climatically-induced fluctuations of episodic erosion and deposition.

3.3 Fan Dynamics

Alluvial fan morphology is a reflection of the changing relationships between aggradation and dissection during fan development. Many fans show a fan-

Photo 3: *Corachos fan, headcut on distal fan surface, looking downfan.*

head trench between apex and midfan locations, within which the channel has a lesser slope than the fan surface. In midfan these two profiles may converge at an intersection point (HOOKE 1968, see also photo 1). In distal locations aggradation may dominate (WASSON 1974). This morhology may simply be the result of progressive ageing (ECKIS 1928) as dissection of the source-area continues and the focus of deposition shifts downfan. However, periods of fanhead trenching will be those of maximum dissection, and as it is the case of the Spanish fans, may be interrupted by episodic aggradation within the fanhead trench.

Slopes of the fan surface and of the channel within the fanhead trench will be adjusted to the prevailing erosional or depositional sediment transport regime at the times of their formation. Hence, a simple model of fan aggradational or dissectional behaviour may be based on the relationships between fan and channel profiles (fig.4) (HARVEY 1987). Fans of the type described above, with an adequate supply of sediment for continued distal aggradation would have profiles of types A or B in fig.4. However, under conditions of limited sediment availability the channel may emerge at the intersection point onto the fan surface and inherit a slope, from there downfan, that was formed under conditions of fan aggradation, a slope steeper than that related to stream processes within the fanhead trench (type C, fig.4). In this situation limited sediment availability prevents modofication of the distal slope by aggradation. At the intersection point, provided that a rapid increase in channel width does not occur, unit stream power may increase (BULL 1979), perhaps sufficiently to cause incision into the fan surface. On crusted fans, where channel widths may be restricted (VAN ARSDALE 1982), intersection-point headcuts may form (type D, fig.4; see also photo 3), resulting in a downfan shift in the position of the intersection point. At each successive intersection point further headcuts may form eventually leading to the total dissection by the channel through the distal fan surface (types E and F, fig.4).

On the Carrascoy fans recent agricultural improvements have obscured the relationships on 4 fans but of the remainder, 4 are of type C, 3 of which are northern mountain-front fans, and 7 are distally dissecting fans of types D and E (fig.4, for examples see fig.3).

The tendency for distal dissection may be expected where the increase in channel slope at the intersection point is greatest, or in other words where there is a marked discrepancy between fan and channel slopes, and where channel width is relatively small (HARVEY 1987). These three morphometric variables can be predicted on the basis of source area and sediment characteristics. Steep fan gradients are associated with small source-areas, especially those generating debris-flow as opposed to fluvial deposits, and especially in lithologies yielding coarse clasts (BLISSENBACH 1952, BULL 1962, HARVEY 1984b). Channel slopes, resulting from fluvial processes, are less steep than fan gradients resulting from debris-flow and sheetflow processes. Channel widths tend to be less on crusted than on non-crusted fans (VAN ARSDALE 1982, HARVEY 1987).

The relationships to drainage area of these three variables for the Carrascoy fans, for which field data are available, are shown in the context of a large sample of Spanish fans in fig.5. The regional relationships have been discussed elsewhere (HARVEY 1987), but for the Carrascoy fans, fan gradients, especially for the northern mountain-front (group 1) fans, tend to be higher than on many other Spanish fans. Of the 15 Carrascoy residuals from the regional regression relationship (fig.5), 11 are positive. A separate regression relationship has been calculated for the Carrascoy fans (fig.5), but with a correlation coefficient of only −0.373 (Standard Error of the etimate: 0.153 log units), it is not statistically significant. This is the only major group of alluvial fans in southeast Spain for which such a relationship is not significant (HARVEY 1987). However when source area drainage-basin relief is also taken into account in a multiple regression analysis the correlation coefficient improves to 0.742 (Standard Error of the estimate: 0.115 log units), and the following relationship is now significant at the 5%-level:

$$G = 0.00009A^{-0.66}B^{1.15}$$

(where G and A are as defined on fig.5 and B is basin relief, m). It is interesting however, that group 1 fans still have strongly positive residuals, which using Student's 't' test, are significantly different at the 5% level from those of groups 2 and 3. This suggests that some factor other than drainage area, basin relief or bedrock geology, which is similar for all three groups, has influenced fan gradient on these northern mountain-front fans. As these are almost the only fans within the whole sample for which there is direct expression in the sediments of tec-

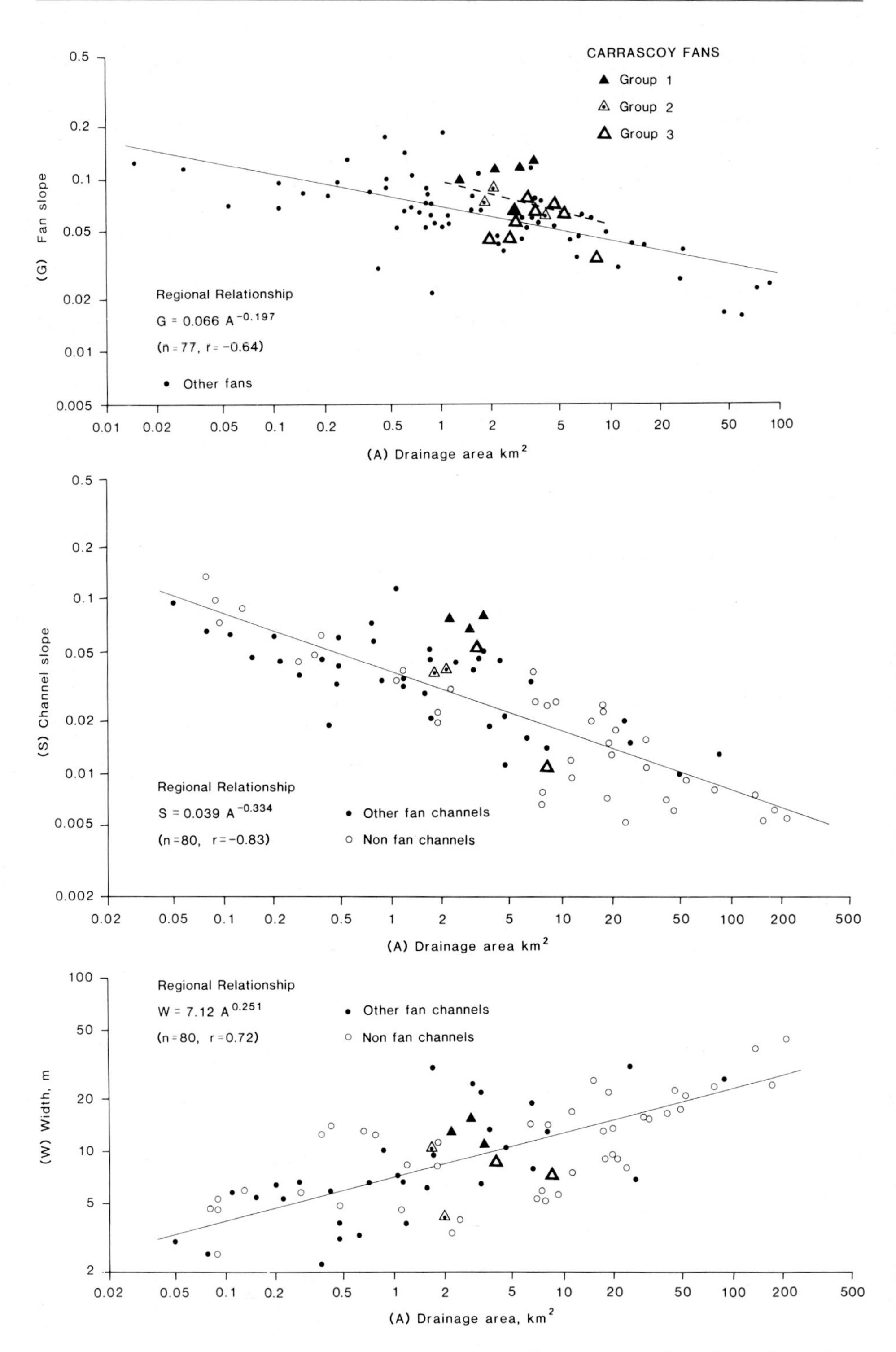

Fig. 5: *Morphometric relationships on Carrascoy fans compared to those throughout southeast Spain (after HARVEY 1987).*

tonic disturbance, and the fan slopes are being under-estimated by the multiple regression equation, the evidence supports tectonic uplift at the mountain front as an influence on fan morphometry.

Channel slopes and widths have been measured on only seven fans, an inadequate number for regression analyses separate from the regional analyses. However for the three group 1 channels measured, channel slopes plot well above the regional trend (fig.5), as might be expected in response to dissection of the steep fan slopes. The others plot near or below the regional trend. All widths cluster within the broad scatter of the regional trend (fig.5).

It was argued above that the midfan trenching tendency, especially apparent on the inter-mountain and southern Carrascoy fans (Groups 2 and 3), would be expected where the discrepancy between fan and channel slopes is greatest. This appears to be born out by the data. Group 1 fans have steep fan gradients and channel slopes, channel slope averaging 0.67 of the equivalent fan slope, whereas Groups 2 and 3, although with lower fan and channel slopes, have a greater discrepancy between the two, channel slopes averaging 0.44 of the equivalent fan slopes.

These trends are illustrated on fig.3, for three examples fans. Roy Fan (group 1) shows intersection-point scour, but no evidence of midfan and distal trenching. Ros Fan (group 2) and Corachos Fan (group 3) both show multiple headcut development in midfan and distal locations and the incipient development of through-fan trenching. In more general terms all Carrascoy fans are currently dissectional with very limited aggradation even on group 1 fans, occuring only in distal locations.

4 Discussion

For the Carrascoy fans the three groups of factors identified earlier; tectonic, climatic and dynamic, have combined to influence the morphological development of the alluvial fans. Regional tectonism has created the gross relief patterns, leading to fan formation, especially at the northern mountain-front. Pedimentation of the mountain-front slopes in the intermontane and southern areas modified the relief prior to fan deposition. Continuing tectonism during fan formation is of limited importance, only influencing the sedimentary sequences of the northern, mountain-front fans and modifying their morphometric relationships.

Fan development relates to a considerable period of the Quaternary with major aggradational episodes apparently responding to climatically-induced periods of high sediment production, and dissectional episodes to periods of low sediment production. These climatically induced alternations in aggradational and dissectional regime are superimposed onto long-term trends that reflect the ageing of the system. These trends are expressed by both the progressive change from fines-rich debris-flow and sheetflood processes to fines-poor, within-channel processes and the change from net aggradation in the earlier phases to net dissection later. The dry, seasonal climates are important not only in their influence on flood sediment production, but also in the development of calcrete-crusted fan surfaces, which in turn influence the fan surface processes.

The mechanisms of fan-trenching, particularly of mid-fan headcut development and distal-fan trenching appear to be threshold controlled (SCHUMM 1977), in a manner similar to that proposed by BULL (1979), related to the threshold of critical stream power. Under conditions of excess sediment supply, unit stream power is inadequate to remove all the sediment supplied from the mountain source areas, resulting in fan aggradation. Under conditions of reduced sediment supply all sediment may be transported to distal locations and trenching may occur in proximal and mid-fan locations. Intersection-point headcut formation and midfan trenching tendencies are enhanced under conditions of reduced sediment-supply, on crusted fans with relatively narrow channels and where the discrepency between fan and channel slopes is great. Because fan slopes relate to past processes there is an inheritance within the system. Such dissection may ultimately lead to the complete trenching of the fan surface and the development of channel continuity from mountain source-area to main drainage, and therefore has implications for the integration of fluvial systems in dry-regions.

The three groups of factors interact in different ways over varying timescales. Over the Quaternary as a whole the regional tectonic patterns, the gross climatic factors influencing sediment availability, and the progressive ageing of the fan morphology may be the most important factors. Over the medium within-Quaternary timescales, climatic fluctuations and their influence on variations in sediment supply may be dominant, but in the shorter term, fan dynamics and the relationships between morphometry and process, control fan behaviour and account for variations between individual fans.

Acknowledgement

I am grateful to the University of Liverpool Staff Research Fund for support towards the costs of the fieldwork and to the staff of the drawing office and photographic sections of the Department of Geography, University of Liverpool for producing the diagrams.

References

ALVARADO, M. (1980): Espagne: In: DERCOURT, J. (Ed.) Geologie de pays europeens, Espagne, Grece, Italie, Portugal, Yougoslavie. 26e Congres Geol. Int. (Paris), 1–54.

AMOR, J.M. & FLORSCHUTZ, F. (1964): Results of the preliminary palynological investigation of samples from a 50 m boring in southern Spain. Bol. R. Soc. Espanola Hist. Nat. (Geol.), **62**, 251–255.

BAKER, V.R. (1977): Stream channel response to floods, with examples from central Texas. Geol. Soc. of Amer. Bull. **88**, 1057–1071.

BIROT, P. & SOLE SABARIS, L. (1959): La morphologie du sudest de l'Espagne. Rev. de Geog. des Pyrennees et du sudouest **30**, 119–184.

BLISSENBACH, E. (1952): Relation of surface angle distribution to particle size distribution on alluvial fans. Journ. Sed. Pet. **22**, 25–28.

BLISSENBACH, E. (1954): Geology of alluvial fans in semi arid regions. Geol. Soc. of Amer. Bull., **65**, 175–190.

BODENHAUSEN, J.W.A. & SIMON, O.J. (1965): On the tectonics of the Sierra de Carrascoy (Province of Murcia, Spain). Geol. en Mijnbouw **44**, 251–253.

BOUSQUET, J.C. (1977): Quaternary strike-slip faults in southeastern Spain. Tectonophysics, **52**, 277–286.

BRUNSDEN, D. & THORNES, J.B. (1979): Landscape sensitivity and change. Inst. Brit. Geogr. Trans New Series, **4**, 463–484.

BULL, W.B. (1962): relations of alluvial fan size and slope to drainage basin size and lithology in wetern Fresno County, California. U.S. Geol. Survey Professional Paper **450B**, 51–53.

BULL, W.B. (1964): Geomorphology of segmental alluvial fans in western Fresno County, California. U.S. Geol. Survey Prof. Paper **352E**, 89–129.

BULL, W.B. (1977): The alluvial fan environment. Prog. in Phys. Geog., **1**, 222–270.

BULL, W.B. (1979): Threshold of critical stream power. Geol. Soc. of Amer. Bull. **90**, 453–464.

BUTZER, K.W. (1964): Climatic-geomorphologic interpretations of Pleistocene sediments in the Eurafrican sub tropics. In: HOWELL, F.C. & BOULIERE, F. (Eds.). African Ecology and Human Evolution. London (Methuen), 1–25.

CUENCA PAYA, C. & WALKER, M.J. (1976): Pleistoceno final y Holoceno en la cuenca del Vinalopo (Alicante). Estudios Geologicos, **32**, 14–104.

DUMAS, M.B. (1969): Glacis et croutes calcaires dans le levant espanol. Assoc. de Geogr. Francais Bull., **375**, 553–561.

DUMAS, M.B. (1977): Le Levant Espagnol, la genese du relief. These de doctorat d'etat, Universite Paris-Val de Marne (Paris XII) CNRS, 520 p.

ECKIS, R. (1928): Alluvial fans of the Cucamunga district, southern California. Journ. Geol., **36**, 225–247.

GEIGER, F. (1970): Dei ariditat in sudostspanien. Stuttgarter Geographische Studien Band **77**, 173 p.

HARVEY, A.M. (1978): Dissected alluvial fans in southeast Spain. CATENA, **5**, 177–211.

HARVEY, A.M. (1984a): Aggradation and dissection sequences on Spanish alluvial fans; influence on morphological development. CATENA, **11**, 289–304.

HARVEY, A.M. (1984b): Debris flows and fluvial deposits in Spanish Quaternary alluvial fans: implications for fan morphology. In: KOSTER, E.H. & STEEL, R.J. (Eds.), Sedimentology of gravels and conglomerates. Can. Soc. Petroleum Geol. Memoir, **10**, 123–132.

HARVEY, A.M. (1987): Alluvial fan dissection: relationships between morphology and sedimentation. In: FROSTICK, L. & REID, I. (Eds.), Desert sediments, Ancient and Modern. Geol. Soc. London, Sp. Publ. **35**, 87–103.

HARVEY, A.M. (in press): Factors influencing Quaternary alluvial fan development in southeast Spain. In: RACHOCKI, A.H. & CHURCH, M.J. (Eds.), Alluvial Fans, a field approach. London (Wiley).

HOOKE, R. le B. (1967): Processes on arid-region alluvial fans. Journ. Geol., **75**, 438–460.

HOOKE, R. le B. (1968): Steady-state relationships on arid region alluvial fans in closed basins. Am. Journ. Sci. **266**, 609–629.

I.G.M.E. (Instituto Geologico y Minero de Espana, 1:50,000 geological maps and accompanying memoirs. Sheets 933 (Alcantarilla), 934 (Murcia), 954 (Totana)).

RIOS, J.M. (1978): The Mediterranean coast of Spain and the Alboran Sea. Ch 1 in: NAIRN, S.C.M., KANES, W.H. & STEHLI, F.G. (Eds.), The Ocean Basins and their margins. Volume **4B**, The Western Mediteranean. New York (Plennum), 1–66.

ROHDENBURG, H. & SABELBERG, U. (1980): Northwestern Sahara margins: terestrial stratigraphy of the Upper Quaternary and some palaeoclimatic implications. In: VAN ZINDEREN BAKKER, E.M. & COETZEE, J.A. (Eds.), Palaeoecology of African and the surrounding Island, **12**, 267–276.

SABELBERG, U. (1977): The stratigraphic record of late Quaternary accumulated series in southwest Morocco and its consequences concerning the pluvial hypothesis. CATENA, **4**, 204–215.

SCHUMM, A.S. (1977): The fluvial system. London, Wiley, 338 p.

VAN ARSDALE, R. (1982): Influence of calcrete on the geometry of arroyos near Buckeye, Arizona. Geol. Soc. of Amer. Bull, **93**, 20–26.

VITA FINZI, C. (1972): Supply of fluvial sediment to the Mediterranean during the last 20,000 years. In: STANLEY, D.J. (Ed.), The Mediterranean sea a natural sedimentation laboratory. Stroudsburg, Pa., 43–46.

WASSON, R.J. (1974): Intersection point deposition on alluvial fans: an Australian example. Geogr. Ann. **56A**, 83–92.

WELLS, S.G. & HARVEY, A.M. (1987): Sedimentologie and geomorphic variations in storm generated alluvial fans, Howgill Fells, Northwest England. Geol. Soc. of Amer. Bull., **98**, 182–198.

Address of author:
A.M. Harvey
Department of Geography
University of Liverpool
P.O. Box 147
Liverpool L69 3BX
England

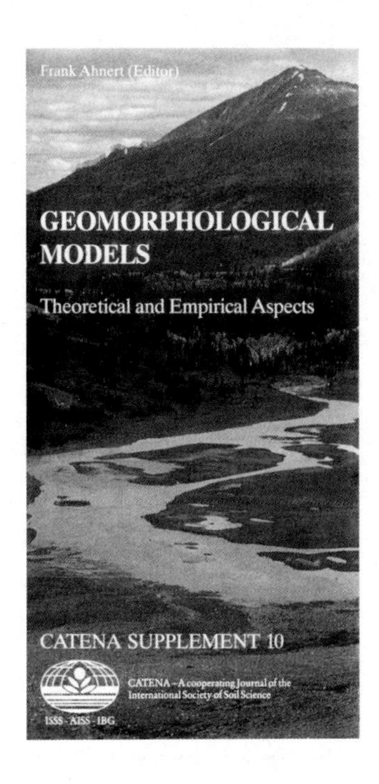

Frank Ahnert (Editor):

GEOMORPHOLOGICAL MODELS

Theoretical and Empirical Aspects

CATENA SUPPLEMENT 10, 1987

Price DM 149,— / US $88.—

ISSN 0722-0723 / ISBN 3-923381-10-7

CONTENTS

CATENA SUPPLEMENT 13 p. 139–146 Braunschweig 1988

A QUANTITATIVE APPROACH TO SCARP RETREAT STARTING FROM TRIANGULAR SLOPE FACETS, CENTRAL EBRO BASIN, SPAIN

C. **Sancho**, M. **Gutiérrez**, J.L. **Peña**, Zaragoza
F. **Burillo**, Teruel

Summary

The existence of four stages of slope evolution that are reflected morphologically by debris-covered triangular slope facets, provides significant data with respect to the development of scarp slopes in the central Ebro Basin (Spain). The deposits from the two most recent stages include archeological remains which allows us to date them as post-Roman (1st Century A.D.) and post-Medieval. A method to calculate the extrapolated segments of these relict landforms is developed. The data supplied indicate a scarp retreat that is calculated to be around 3 m per 1000 years.

Resumen

La existencia de cuatro etapas de evolución de vertientes que quedan reflejadas morfológicamente por facetas triangulares de vertiente coronadas por detritus, proporciona datos significativos de cara al desarrollo de estas laderas. Los depósitos de las dos etapas más recientes presentan restos arqueológicos que permiten su datación como postromanas (S. I d. de C.) y postmedievales. Se desarrolla un método para el cálculo de los segmentos extrapolados de estas formas relictas. Los datos suministrados indican un retroceso del escarpe que se estima en 3 m cada mil años.

ISSN 0722-0723
ISBN 3-923381-13-1

3-923381-13-1/88/5011851/US$ 2.00 + 0.25

1 Introduction

The presence of triangular slope facets is characteristic of slope evolution in periglacial and semiarid areas (BÜDEL 1970, 1982, EVERARD 1963, GERSON 1982, GOSSMAN 1976), even though these landforms are not extremely common. This relict morphology is studied in very few works. Research into these landforms provides valuable information in relation to the retreat of old escarpments. The existence of several formative stages of triangular slope facets allows us to distingiush relative rates of scarp recession. Moreover, it is also possible, from the data obtained from deposits found on the slopes, to make quantitative calculations of the rate of scarp retreat. However, a serious problem arises when one attempts to interpret the causes bringing about the alternation between

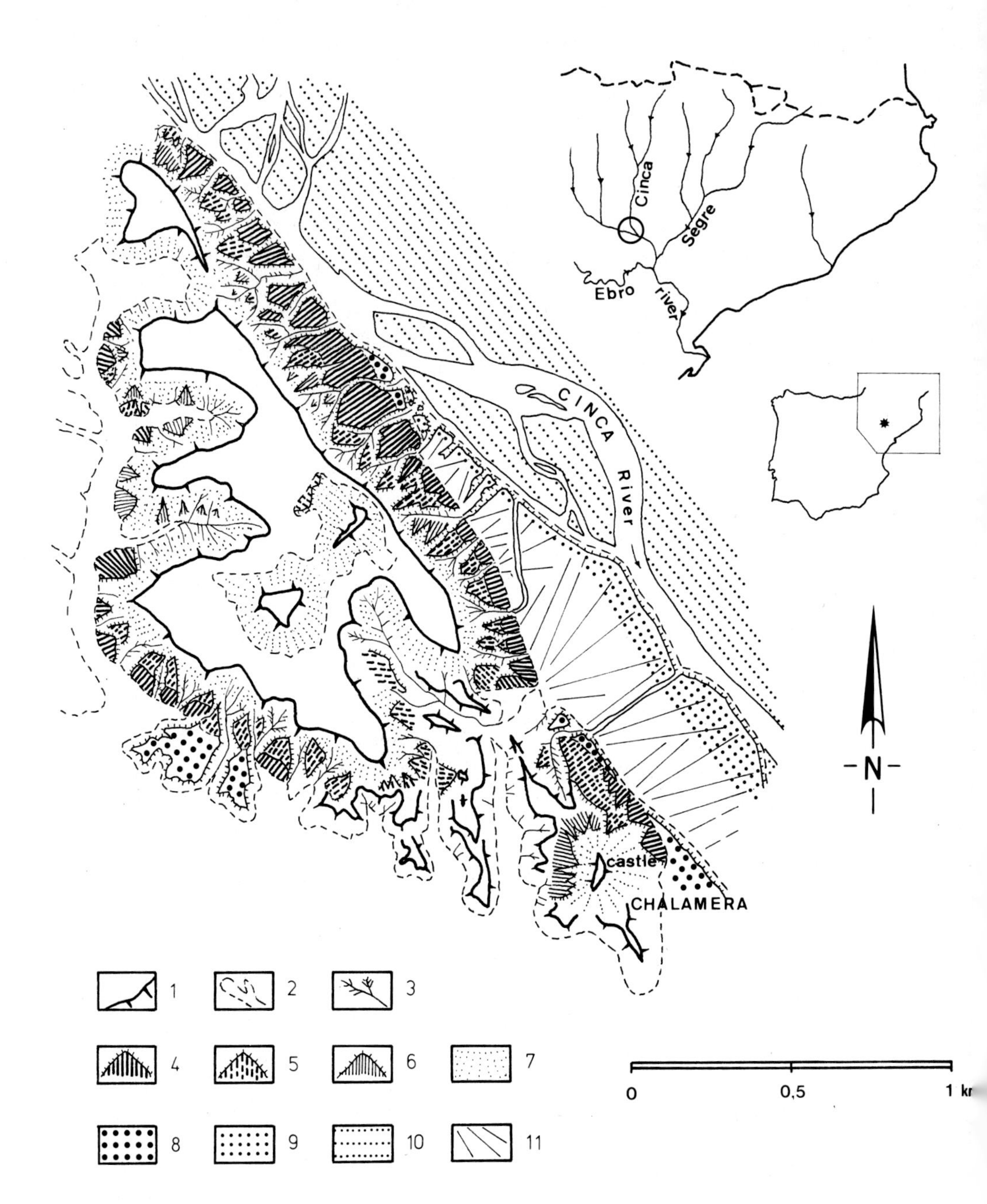

Fig. 1: *Location and geomorphological map of the Chalamera area (Huesca province).*

1: Mesa and structural scarp. 2: Flat bottomed valleys. 3: Rills and gullies. 4,5,6 and 7: Slopes facets (S_4, S_3, S_2, s_1). 8, 9 and 10: Terraces (T_3, T_2, T_1). 11: Alluvial fans.

accumulation and erosive stages which have given rise to the successive triangular slope facet stages.

2 The Study Area

The slopes under study are found within the Ebro Basin, Northern Spain, developed in Tertiary rocks, near the confluence of the Alcanadere and Cinca rivers, near to Chalamera, Huesca (fig.1). The bedrock materials are horizontal and comprise shale with intercalations of sandstone and limestones corresponding to a mudflat environment and which form part of a Miocene alluvial fan system. The area is some 200–300 metres above sea level and it possesses continental Mediterranean climatic characteristics with total annual rainfall being approximately 400 mm, extremely dry summers and the average annual temperature around 15°C (mean annual temperature range = 20°C). In geomorphological terms we are talking of a mesa topped by limestones, resulting from the downcutting of the Alcanadre-Cinca fluvial system, which has developed different stepped Quaternary terraces. Between the mesa top and these terrace systems several phases of slope development can be seen.

3 Triangular Slope Facets

Four stages in the retreat of the escarpment are reflected in four separate accumulative relict landforms (labelled S_1–S_4, youngest to oldest, on fig.1) on the Chalamera mesa slopes. These slope remnants are termed **talus flatirons** (KOONS 1955); **flatirons** (EVERARD 1963); **Dreieckshänge** (WIRTHMANN 1964, BÜDEL 1970); **Versants tripartites** (GOSSMAN 1976); **tripartites slopes** and **triangular slope facets** (BÜDEL 1982).

These triangular slope facets are rare in the Ebro Basin, and when they do occur, usually only a single facet with little lateral continuity is to be seen. In this particular area the number of facets is really unusual. They lie parallel to the mesa's present scarp, with the oldest facets being furthest from the scarp, indicating a parallel retreat of its edges and therefore a progressive reduction in size of the mesa (fig.2).

The morphology of these triangular facets is that of slopes with steep apices and orientated towards the scarp (photo 1). The slopes are topped with debris that generally comprises flat parallel-piped limestone blocks, orientated towards the greatest slope and embedded in a sandy-shale matrix.

The facet profiles are smooth and concave as those described by BÜDEL (1970). The maximum slope angles, obtained from the arithmetical average of the measurements taken on the upper part of all the facets are the following: $S_4 = 16°$; $S_3 = 22°$; and $S_2 = 26°$. the values of the minimum angles corresponding to the lower part of the concavity vary between 22° and 5°. The figures for the maximum angles reflect a decrease in value away from the escarpment, which can be interpreted as due to greater erosion of the upper parts of the older profiles. In some cases, mass wasting on a more recent slope can fossilize the apical areas of the older facets, with the profile alone remaining and with, therefore, the facet morphology vanishing.

The slope profile starting from the facet apex and running towards the scarp is also concave (fig.2) and the values of the maximum angles of this concavity

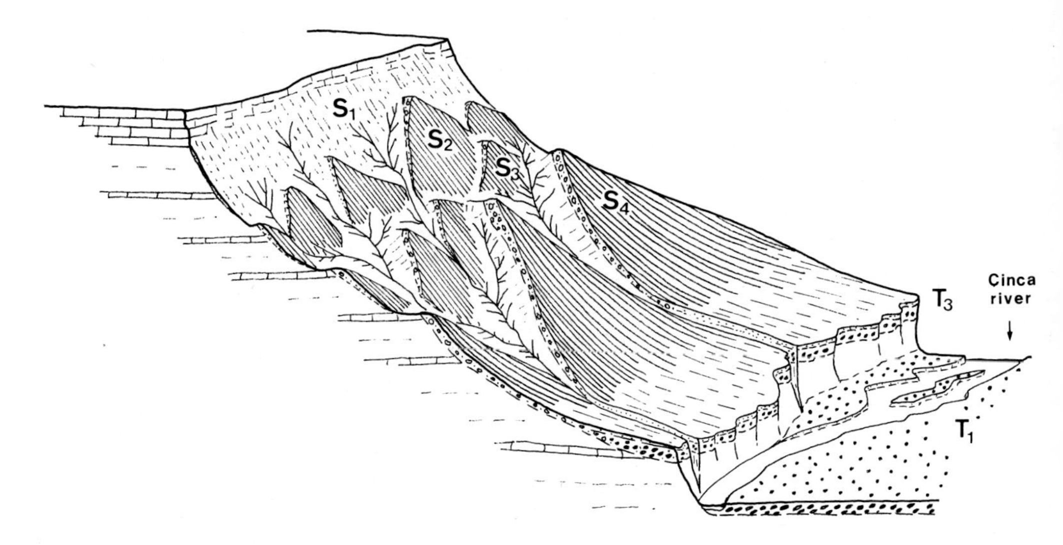

Fig. 2: *Idealised block-diagram of the four stages of slope evolution, in which three phases of triangular slope facets are observed.*

Photo 1: *Triangular slope facets and mesa scarp.*

fluctuate between 5° and 41°, the cause here being variation of bedrock lithology and of the degree of downcutting of the rills and gullies. There are also some points at which these concavities of reverse slope do not exist, and which correspond to more recent slope profiles that cut off the upper parts of the older slopes.

The most recent slope (S_1) likewise presents a continuous and regular concave profile and only rarely is a small convex element present, related to resistent rocks on the upper part of the slope profile. The average for the angles measured for the concave unit is 5° for the lower parts and 34° for the top ones. Around Chalamera castle incipient tripartite slopes can be seen that have been produced as the result of rapid downcutting by modern gullies, although, on the whole, debris covered slopes with more or less parallel gullies are dominant. In connection with the gullying, piping processes are evident which affect both the slope deposits and the Miocene bedrock. The piping is more intense in the alluvial fans that exist at the foot of the slopes; circumstances that are quite common in the Ebro Basin (GUTIERREZ et al. 1987).

The facets comprise relict slope accumulations and their separation is caused by gully incision in midslope, where there is a confluence of tributary rills. The gullies, once having cut the facets, run down to the low gently-sloping areas and it is there where they deposit their load in the form of alluvial fans. Once the tripartite slopes have been isolated a break in the denudation process takes place (BÜDEL 1970) and each facet behaves as a relict relief, functioning quite independently of the upper parts of the slopes. The slope accumulation processes, followed by a cutting stage, with consequent isolation of the facets, have taken place four times within the most recent evolution of the Chalamera mesa slopes.

The oldest slope deposits interfinger laterally with the River Cinca's 20 m terrace. The materials on the two youngest slopes (S_2 and S_1) contain a great deal of archeological remains, primarily pottery, especially around Chalamera Castle hill, which has been both constantly and heavily populated (DOMINGUEZ 1975, DOMINGUEZ et al. 1984). In the S_2 slope deposits there have occasionally been found pieces of handmade pottery dating from the Early Iron Age, others from the Iberian period (3rd to 1st centuries B.C.), which are in fact the most abundant, and some Late Iberian with A and B Campanian and terra sigillata, the typology of which indicates its existence until the Roman Empire period (end of the 1st century A.D.). All these pieces of pottery are found within the slope deposits and so they indicate dates **postquem** for these deposits. The archeological remains within the S_1 slope debris include all those mentioned for the older slope together with others from the Medieval Period.

4 The Calculation of Scarp Retreat

In order to get an approximate idea of the retreat rate for the successive scarps existing at each stage, which is indicated by a tripartite slope, the profiles of 27 facets have been surveyed, and these have been represented graphically by means of a coordinated reference system comprised of the values of the horizontal distance (abscisa axis) measured with calipers on aerial photographs (ap-

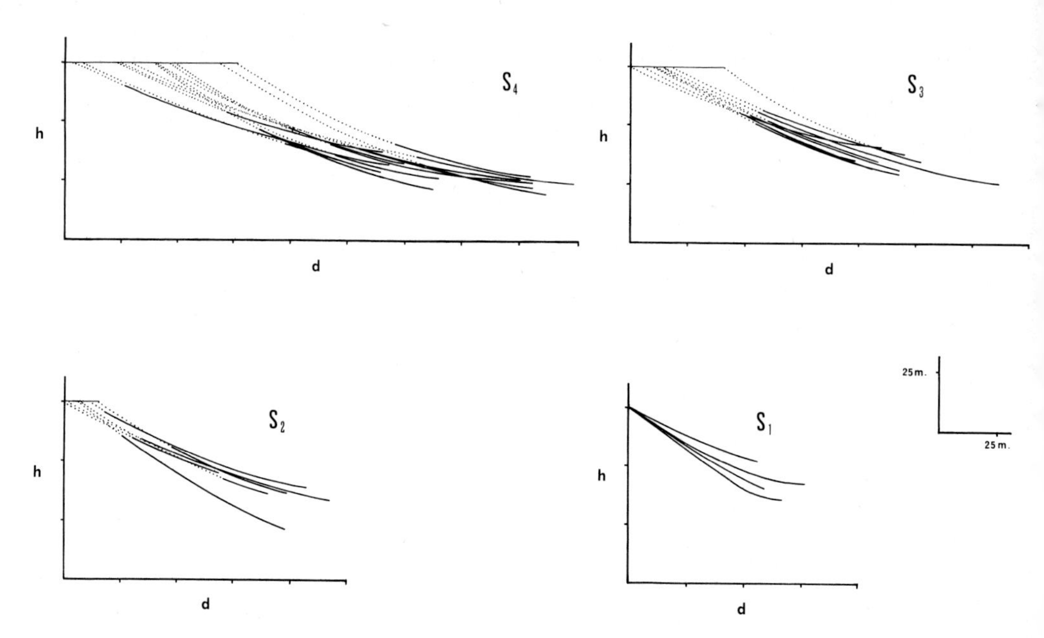

Fig. 3: *Profiles of of triangular slope facets (unbroken line) and extrapolated curves (broken line).*

h: height of triangular facets apex with respect to present scarp. d: distance from the facet apex to the present scarp (measured from aerial photographs).

prox. scale 1:18.000) and of the height (ordinate axis) calculated with a Thommen 3B4 altimeter (error margin of ± 1 m) between the scarp edge and the facet apex (fig.3). Of these two coordinates the one corresponding to the height is a fixed value; whereas the horizontal distance with respect to the scarp is variable, since the scarp is not, nor has been, rectilinear. The resulting curves indicate an evident dispersion that is due to scarp inflections at different stages of development, which must correspond to different retreat rates. Therefore, the construction of an average curve for each of the stages by statistical means does not afford any reliable data, since the distances that exist with respect to the scarp at each stage are both variable and unknown. The method used to obtain estimates of scarp retreat consists of extrapolating all the curves of the profiles represented. The extrapolation was done by means of the Statworks programme on a Macintosh 512 Kv computer, which allowed us to achieve a polynomial regression, and by calculating the coefficients of the equation $y = a+bx+cx^2$, since the fit is good. From the intersection of this curve with the horizontal straight line, the prolongation of the mesa's plane, one obtains the scarp position for stages S_2, S_3 and S_4. The relative differences between one position and another make it possible to obtain the different scarp retreat rates between successive stages.

In order to calibrate the retreat rates

it is necessary to estimate the ages of the facets by making use of the archeological information derived from the deposits found in the facets. We have already shown that the youngest slope (S_1) dates, via its archeological remains, from the post-Medieval period, and that the remains found on the newest tripartite slope (S_2) correspond to the post-Roman era (1st century A.D.). Although these relative datings do not give exact figures, what they do establish is the maximum time span between the two slopes, from which minimum scarp retreat rates can be obtained. Calculations indicate a mean rate of about 3 m per 1000 years (3,000 Bubnoff units). The true values are probably greater because, as has already been shown, the youngest slope appears to be in the early stages of slope destruction, by a generalised downcutting process by rills and gullies, following an earlier period of debris-accumulation. We do not possess data, however, relating to the period during which this slope changed from an accumulational to a degradational regime.

By extrapolation from this estimate for the retreat rate for slope S_2, approximate ages of the two older facets are estimated to be: Slope S_3 circa 5000 years, and S_4 circa 12000 years. The interfingering of the oldest slope's materials with terrace deposits allows us suggest an approximate date (late Würm) for the latter by this extrapolation.

These figures accord with similar estimates made by other authors. Thus, SCHUMM & CHORLEY (1966), in YOUNG (1974), when referring to lithologies and morphoclimatic areas similar to the one under study here, give scarp retreat rates of 2,000–13,000 B units. SAUNDERS & YOUNG (1983) give the average retreat rate a value of 3,000–3,500 B units, for semi-arid environments and steep slopes.

5 Conclusions

The presence of four slope evolution stages, the three oldest being regarded as tripartite slopes, allows us to obtain data on their development. the existence of these triangular slope facets is due to changes in the morpho-dynamics. The deposits on these slopes appear to accumulate under relatively cold environmental conditions and the incisions appear to occur in response to climatic conditions similar to those of the present day (BURILLO et al. 1986). For the two latest stages, anthropogenic change may have played at least a contributive role in setting off the downcutting process. These changes cause a successive scarp retreat and a reduction in the area occupied by the mesa.

The complexity of the unusual sequence that appears at the confluence of the Cinca and Alcanadre rivers, in the Chalamera region, in which four evolution slope stages can be seen, indicated by triangular facets, from late Würm (?) until the post-Medieval priod, indicates the sensitivity of erosion/deposition thresholds on these slopes.

Acknowledgement

We should like to thank Professor E. Rubio and Dr. E. Dolado of the Dept. of Biostatistics of the University of Zaragoza for all their help with the mathematical part of this work.

References

BÜDEL, J. (1970): Pedimente, Rumpfflächen und Rückland-Steilhänge. Z. Geomorph. N.F., **14**, 1–57.

BÜDEL, J. (1982): Climatic geomorphology. Princeton University Press. 443 p.

BURILLO, F., GUTIERREZ, M., PEÑA, J.L. & SANCHO, C. (1986): Geomorphological processes as indicators of climatic changes during the Holocene in the North-East Spain. In: LOPEZ VERA, F. (Ed.), Quaternary Climate in the Western Mediterranean. Proceedings of the Symposium on climate fluctuations during the Quaternary in the Western Mediterranean Regions. Madrid. 31–44.

DOMINGUEZ, A. (1975): Nuevos hallazgos arqueológicos en Chalamera (Huesca). Miscelánea Arqueológica Profesor Antonio Beltrán. Zaragoza, 187–195.

DOMINGUEZ, A., MAGALLON, M.P. & CASADO, M.P. (1984): Carta arqueológica de España. Huesca. Diputación Provincial de Huesca. 288 p.

EVERARD, C.E. (1963): Contrast in the form and evolution of hillside slopes in Central Cyprus. Inst. British Geogr. Trans. **32**, 31–47.

GERSON, R. (1982): Talus relicts in deserts: A key to major climatic fluctuations. Israel J. of Earth Sci., **31**, 123–132.

GOSSMANN, H. (1976): L'impotance des processus se déroulant à la ligne de partage locale des eaux pour l'évolution des versants sous la dominance du ruissellement pluvial (à l'aide de formules mathématiques élémentaires). Actes du Symposium sur les versants en pays méditerranéens. C.E.G.E.R.M., V. 139–143. Université d'Aix-Marseille II.

GUTIERREZ, M., RODRIGUEZ, J. & BENITO, C. (1987): Piping processes in badland areas, Middle Ebro Basin, Spain. This volume.

KOONS, D. (1955): Cliff retreat in the southwestern United States. Amer. J. Sci., **253**, 44–52.

SAUNDERS, J. & YOUNG, A. (1983): Rates of surface processes on slopes, slope retreat and denudation. Earth surface processes and landforms, **8**, 473–501.

SCHUMM, S.A. & CHORLEY, R.J. (1966): Talus weathering and scarp recessions in the Colorado Plateau. Z. Geomorph. N.F., **10**, 11–86.

WIRTHMANN, A. (1964): Die Landformen der Edge-Inseö in Südost-Spitzbergen. Ergebnisse der Stanforland-Expedition, V. **2**, 53 p.

YOUNG, A. (1974): The rate of slope retreat. Inst. British Geogr., Special Publ., **7**, 65–78.

Addresses of authors:
C. Sancho, M. Gutíerrez
Departamanto de Geomorfología y Geotectónica
Facultad de Ciencias
Universidad de Zaragoza
50009 Zaragoza, Spain
J.L. Peña
Departamento de Geografía General
Facultad de Filosofía y Letras
Universidad de Zaragoza
50009 Zaragoza, Spain
F. Burillo
Departamento de Prehistoria
Colegio Universitario de Teruel
Teruel, Spain

CATENA SUPPLEMENT 13 p. 147–164 Braunschweig 1988

THE GEOMORPHIC SIGNIFICANCE OF PROCESS MEASUREMENTS IN AN ANCIENT LANDSCAPE

A.J. **Conacher**, Nedlands

Summary

Southwestern Australia is a low relief, low altitude plateau underlain by Archaean granites and gneisses. Previous climates have left a legacy of deep-weathered, duricrusted residuals, saline playa chains, and sandplains, currently being modified under a mostly semi-arid, mediterranean climatic regime and significantly affected by human activities. Geomorphic questions include the reasons for the differential stripping of the deep-weathered soils, the origins of the playa chains and sandplains, and the relative importance of overland flow, throughflow and wind as agents of landform development. This paper considers whether measurements of contemporary geomorphic processes are able to contribute usefully to understanding past and present landform development.

Problems with process measurements include the apparently changing seasonality of the climate and the occurrence of unpredictable storm events associated with tropical cyclones and thunderstorms. Soil infiltration rates vary considerably between the dry summers and the wet winters. Spatial sampling of the total landsurface to provide representative process data is an unrealistic ideal. Nevertheless, process measurements do provide some information on contemporary rates of change of the southwestern Australian landsurface. The most promising area for future process research to further understanding of present-day landform development is the relationship between slope and stream processes.

Much more problematic is the relevance of contemporary process measurements to past events and processes. Here the most useful line of research would appear to be the geomorphic implications of the relationships between sedimentary bodies in the landscape and the processes by which they were deposited.

Resumen

El sudoeste de Australia está constituído por una plataforma de poca altitud y relieve, formada por granitos y gneises arcaicos. Los climas pasados han dado lugar a resíduos profundamente meteorizados y encostrados, a cadenas de playa salina y a llanuras de arena, todo lo cual es actualmente modificado bajo condiciones de un clima de régimen mediterráneo, esencialmente semi-árido, al mismo tiempo que se ve afectado de forma notable por las actividades humanas. Los problemas geomorfológicos

ISSN 0722-0723
ISBN 3-923381-13-1

3-923381-13-1/88/5011851/US$ 2.00 + 0.25

que se plantean en el área incluyen la aclaración del motivo por el que se produce una erosión diferencial de los suelos profundamente meteorizados, los orígenes de las cadenas de playa y las llanuras arenosas, y la importancia relativa de la escorrentía superficial y subsuperficial y la del viento como agentes del desarrollo de las formas del relieve.

Aunque un muestreo eficaz de los distintos paisajes para obtener datos sobre los procesos representativos es un ideal poco realista, la medición de procesos proporciona sin embargo una interesante información sobre las tasas de la evolución de la superficie del terreno del sudoeste australiano. El tema más prometedor cara a una futura investigación sobre procesos para entender mejor el desarrollo actual de las formas del relieve es el de dilucidar las relaciones entre procesos en las vertientes y procesos fluviales. La importancia de la medición de los procesos actuales en la explicación de los eventos pasados no es clara, pero sin embargo en este caso el tema de investigación que parece más apropiado sería el de las implicaciones de las relaciones entre conjuntos sedimentarios en el paisaje y los procesos mediante los que fueron depositados.

1 Introduction

The aim of this paper is to consider whether measurements of contemporary processes in southwestern Australia can be of geomorphic significance. In raising the question, it is assumed that the rationale for making contemporary geomorphic process measurements is to elucidate the ways in which landforms have developed in the past, and are continuing to change, and not merely to identify rates and mechanisms of movement of materials for their own sake.

To place the discussion in context, the current 'Mediterranean' climate and landscape are briefly described. This permits some of the geomorphic questions to be identified. The paper then focusses on process measurement studies and related research that have been carried out in the region, and on problems of undertaking such research, with reference to the specific geomorphic questions identified. Coastal processes are not considered.

2 Climate and Landscape of Southwestern Australia: Some Geomorphic Questions

Southwestern Australian summers are dominated by the descending arm of the Hadley cell. The associated subtropical high pressure belt leads to a flow of dry, warm air from the east, and clear skies. Occasional southern incursions of air masses associated with tropical cyclones disrupt this pattern. In winter, the dry subtropical ridge covers the northern parts of Western Australia while westerly eddies, comprising depressions and anticyclones embedded in the global, westerly airstream, provide the cool wet season for the southern areas. Mean annual rainfall decreases from 1400 mm over the west coast ranges to less than 300 mm inland (fig.1), with more than three times as much rain falling in winter (May–October) than in summer (November–April) (SOUTHERN 1979). Summers are hot and dry—especially inland—and annual evaporation rates considerably exceed mean annual rainfall. Frosts occur regularly in winter, in inland locations, but snowfalls are extremely rare (tab.1).

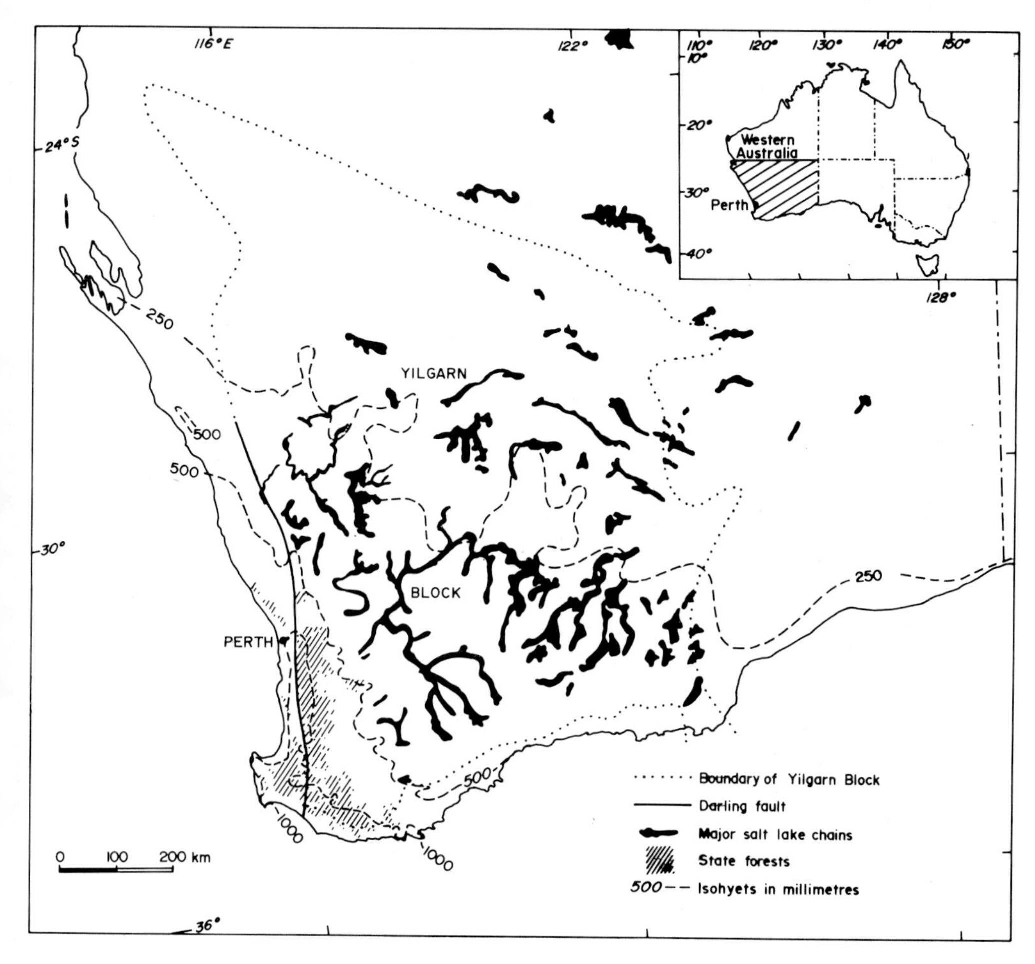

Fig. 1: *Southwestern Australia: showing Yilgarn block, major salt lakes, State forests, Darling fault and key rainfall isohyets.*

Sources: PRIDER 1977, 27, FORESTS DEPARTMENT 1982, MULCAHY 1973, Fig.2, AUSTRALIAN BUREAU OF STATISTICS WESTERN AUSTRALIAN OFFICE 1977, 45.

Past climates influencing southwestern Australia have reflected northerly or southerly shifts of the Hadley cell and the westerly airstream (WYRWOLL & MILTON 1976, WYRWOLL 1979), global cooling and warming periods, and latitudinal shifts of the Australian tectonic plate (BROWN et al. 1968). The narrowness of the continental shelf in the southwest means that sea level changes have had relatively little **direct** climatic effect except along the immediate coastline. Glaciation has not occurred since the Permian.

The land mass under these climatic influences is essentially the 'Yilgarn block' of JUTSON (1934) (fig.1), a shield landscape consisting primarily of granites and gneisses which have been dated radiometrically at between about 2600 and

	Southern Cross	Northam	Katanning
RAINFALL			
mean (mm)	281	435	491
highest one day (mm)	84	126	126
wet days - average no.	69	93	115
TEMPERATURE			
mean max (degrees C)	25.7	25.1	22.1
mean Jan. max	34.7	33.9	30.3
mean min	10.4	10.9	9.2
no. days 30 degrees and over	121.8	110.0	60.5
no. days 40 degrees and over	11.2	11.4	1.9
no. days 2 degrees and under	30.8	17.7	18.0
EVAPORATION			
annual mean in mm (approx)	2200	2000	1300
EFFECTIVE RAINFALL			
no. months per year during which average rainfall exceeds effective rainfall	3	5	6

Tab. 1: *Selected climatic data for three wheatbelt towns: for locations refer fig.3.*

Source: AUSTRALIAN BUREAU OF STATISTICS WESTERN AUSTRALIAN OFFICE 1977, pp. 42, 48–49, 51 and 53.

3100 million years, with basic Early and Late Proterozoic intrusives. Other than some minor occurrences of Cretaceous basalts in the south there is no history of vulcanism since the Proterozoic. The drainage systems, and therefore probably the relief, had become very subdued by the Late Cretaceous. A major period of deep weathering and lateritisation took place in the Oligocene and/or Miocene, but since the Middle Miocene, lateritisation is thought to have occurred intermittently and only in coastal areas of the southwest that have a moderate and strongly seasonal rainfall. The shield was uplifted by about 300 m, probably in the Pliocene, and is now a gently undulating plateau with a mean altitude of 300–450 m above sea level. Subsequent tectonic activity has been relatively insignificant (JOHNSTONE et al. 1973, PRIDER 1977, McARTHUR & BETTENAY 1979).

To the south the plateau slopes gently to the Southern Ocean, where the Australian tectonic plate separated from Antarctica about 60 million years ago. To the west the plateau terminates abruptly at the 1000 km long Darling Fault Scarp which runs parallel to the coast a few tens of kilometres inland (fig.1). There have been up to 15,000 m of vertical movements since the Permian and a history of uplift since at least the Proterozoic. The fault represents the boundary between the Australian and Indian tectonic plates which separated some 150 million years BP (JOHNSTONE et al. 1973).

The western, higher rainfall part of the Yilgarn block is characterised by an extensive, bauxitic plateau surface weath-

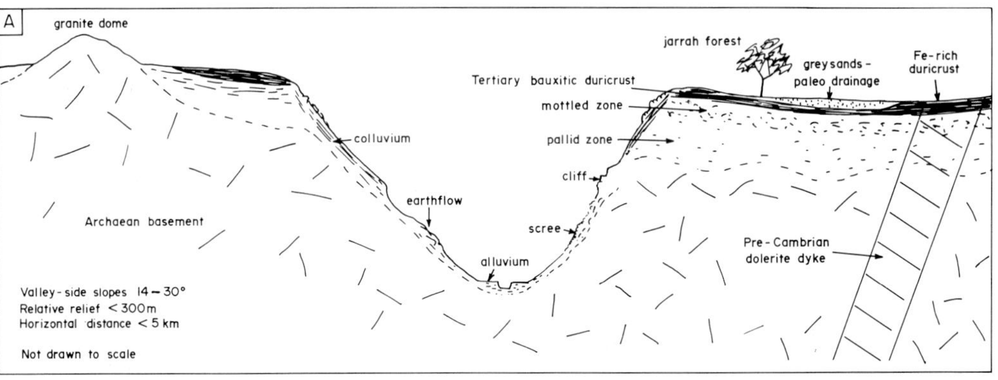

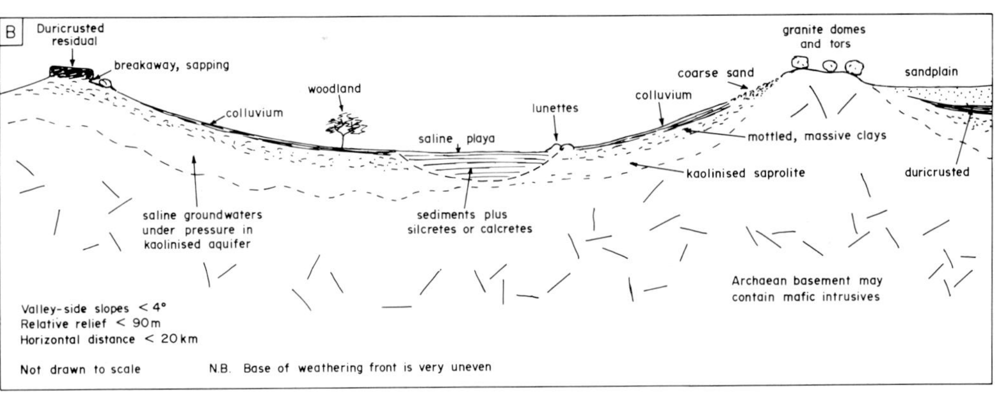

Fig. 2: *Diagrammatic cross sections through southwestern Australian valleys: A in the Darling Ranges, immediately east of the Darling Fault (fig.1), and B in the central Yilgarn Block.*

Source: field observations

ered to depths of up to 50 m. However, the plateau is broken by occasional granitic hills and relatively deeply incised (up to 300 m), steep-sided valleys which debouch through the Darling Scarp on to the coastal plain (BETTENAY & MULCAHY 1972); fig.2A.

In contrast, maximum relative relief in the inland, drier (<500 mm mean annual rainfall) plateau is only 90 m over several kilometres, and the broad (up to 15 km wide), low-gradient (<4 degrees) valleys often contain chains of saline playas (fig.1) with fringing lunettes (BETTENAY 1962). The playa chains are thought to be a relic of the Late Jurassic-Early Cretaceous drainage, which subsequently has had only a relatively minor sedimentary history (JOHNSTONE et al. 1973). Lateritic duricrusts are much less extensive than the bauxitic duricrusts to the west, occurring as relatively minor, steep-sided residuals on the interfluves between the valley systems (fig.2B). The abrupt edges of the duricrust residuals are aptly termed 'breakaways'. Kaolinised, deep-weathered soils are still characteristic, with profiles generally deepest in mid valley-side locations. Siliceous hardpans and, in drier areas, calcretes, underlie some valley floors. Extensive 'sandplains' occur in several areas (MULCAHY 1971, BREWER & BETTENAY 1973, PILGRIM 1979, McARTHUR & BETTENAY 1979).

Four questions may be identified for which contemporary process measurements may be of relevance to understanding the geomorphic development of the landscapes described above. These are as follows:

a) Why has denudation of the drier, inland areas been more extensive than that of the wetter, western areas (fig.2)?

b) What are the relative contributions of wind, overland flow and throughflow to landform development?

c) Are the saline playas responses to paleo-environments or are they currently in the process of formation?

d) What is the origin of the sandplains?

3 The Contribution of Process Measurements to the Geomorphic Questions

3.1 Measurements of Surface Wash

Working in the Wungong (figs.2A and 3), one of the western, deeply-incised valleys, PUVANESWARAN (1981) measured overland flow and sediment movement on ten experimental plots located on landsurface unit 5 of the nine unit landsurface model (CONACHER & DALRYMPLE 1977). Long-term erosion rates were estimated and related to rates of valley widening, using the bauxitic duricrusts as a marker bed (PUVANESWARAN & CONACHER 1983). Valley-widening was estimated to have occurred at rates of the order of one metre per one million years (range 0.2–3.0 m) over the past 30–40 million years.

In a drier (about 400 mm mean annual rainfall) inland area east of Narrogin (figs.2B and 3), PILGRIM (1981) carried out similar measurements on landsurface units 5 and 6. Twelve plots were located on three catenas, all under relatively undisturbed woodland. A continuous record of slope sediment movement over a period of over 6 years was obtained. Although this sediment movement record permitted estimation of rates of slope development with a greater degree of confidence than was possible using PUVANESWARAN's data (PILGRIM et al. 1986), it was not feasible to check PILGRIM's findings with a stratigraphic marker bed. Nevertheless, it was estimated that slope change on landsurface unit 5 was occurring at a rate of the order of 10 m per one million years (range 8.4–24.4 m), possibly explaining the greater lateral extent of stripping of the duricrusted surface in the drier parts of the Yilgarn block.

Some of the sediments from the stripped landsurface have been retained in the inland playas and the wide, flat, valley systems; most have been removed by fluvial processes and redeposited in the thick sediments west of the Darling scarp (JOHNSTONE et al. 1973).

3.2 Measurements of Stream Sediment Discharge

Streamflow monitoring of a few forested catchments by officers of the Public Works Department—mainly to provide

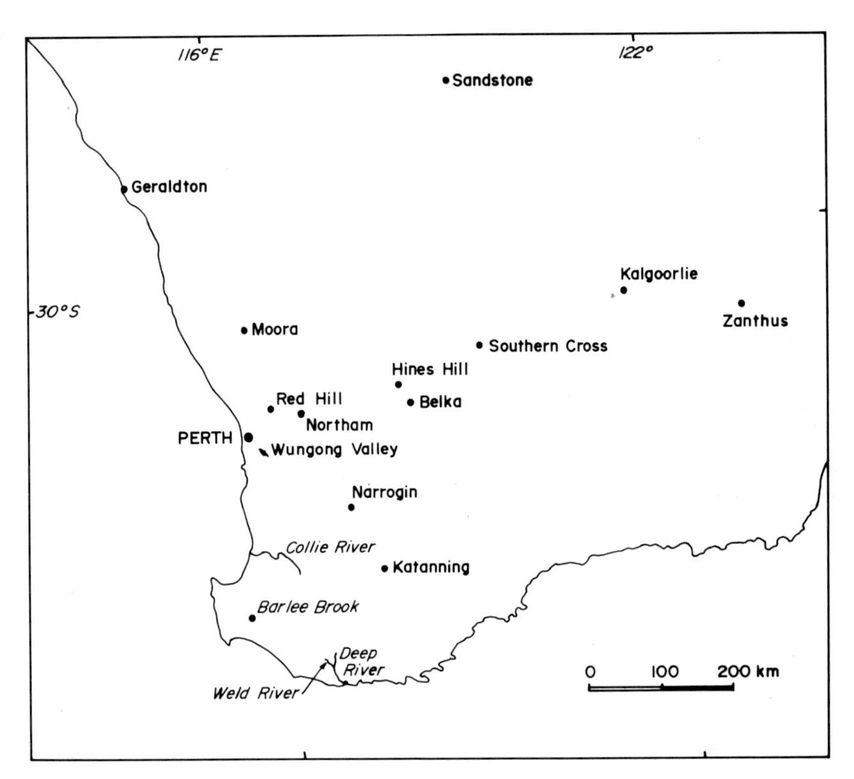

Fig. 3: *Locations of places referred to in the text.*

data on the effects of intensive forestry practices on sediment production in the extreme southwest—has produced some data on natural rates of sediment movement in streams (STEERING COMMITTEE 1980). In all cases sediment yield is very small, providing support for the very low rates of natural erosion identified by PUVANESWARAN (1981).

Fine suspended sediment (<0.63 mm diameter) in four streams, three with no permanent clearing in their catchments and with 2.5% cleared in the fourth, was estimated to comprise 80% of the total suspended sediment load. Measured at four to six week intervals for three years (1976–78), mean annual flow-weighted fine sediment concentrations ranged from 4.5 to 24.4 mg l^{-1}. Upper estimates of annual sediment loads per unit area of the four catchments ranged from 300 to 7100 kg·km^{-2}. The highest sediment discharge occurred in the Weld River (fig.3), where 82% of the annual sediment load for 1978 was transported during a four-day flood event. In this catchment 4% of the forest had been clear-felled. The two catchments with no clear-felling but subject to selective logging yielded upper etimates of annual sediment loads per unit area of 300 to 800 kg·km^{-2}. The bulk densities of these sediments are not known, but applying the mean (1.61) of the bulk densities determined by LOUGHRAN in PILGRIM et al. (1986) yields a surface lowering rate of 0.2 to 0.5 m per one million years for the latter two catchments (Deep River and Barlee Brook: fig.3). Problems with making such extrapolations are discussed below under 'problems of scale'.

3.3 Measurements of Subsurface Water

There is considerable research being undertaken in southwestern Australia by State government agencies, the Commonwealth Scientific and Industrial Research Organisation (CSIRO) and researchers from tertiary educational institutions, on the severe problems of secondary soil and water salinisation. Most of this research is being carried out in the wetter, forested areas in order to monitor the effects of bauxite mining, intensive forstry and agricultural practices on stream water quality (PECK & HURLE 1973, SHEA & HATCH 1976, BATINI & SELKIRK 1978, STEERING COMMITTEE 1978, 1980). An increasing amount of monitoring is also being carried out in the drier, inland agricultural areas in relation to the problem of secondary soil salinisation (BETTENAY et al. 1964, PECK 1978, WILLIAMSON & BETTENAY 1979, CONACHER 1982, PECK et al. 1983, CONACHER et al. 1983).

Although some of this work enables the geomorphologist to identify natural rates of salt flow, there are few analyses of the ionic composition of the salts and the data are therefore of limited value for interpreting catchment geochemistry and rates of denudation by solution processes (cf. CLEAVES et al. 1970). Those analyses which have been carried out (for example by HAND 1974, HINGSTON & GALAITIS 1976, STEERING COMMITTEE 1980) have found that in many locations, particularly towards the west, the composition of the soluble salts closely resembles that of sea water. This has led to the conclusion that the salts are mostly cyclic in origin; derived from the sea and deposited in rainwater. Thus the contribution of solution processes to landsurface denudation in the western part of this region appears to be small.

Nevertheless, subsurface water movements have been shown to be hydrologically and geomorphologically significant. Monitoring by the Public Works Department of paired catchments in the Collie drainage basin (fig.3) in 1980/81 found that, in the forested catchment, throughflow accounted for 91% of streamflow, deep groundwater 7% and surface runoff 2% (STOKES & LOH, pers. comm. 1982). In the cleared, paired catchment, the composition of the increased streamflow changed to throughflow - 56%; deep groundwater - 24%, and surface runoff - 20%. Near Moora, SHERWOOD (1969) showed with dye-tracing that throughflow is an important agent responsible for the headwards extension of first-order streams (see also CONACHER 1975). Subsurface water undoubtedly contributes to the sapping and retreat of breakaways (photo 1), and on some valley-side slopes in wetter areas throughflow also contributes to rapid mass movements (PILGRIM & CONACHER 1974).

BETTENAY et al. (1964) identified iron and aluminium in solution in the groundwaters present in the 'pallid zone' of the inland, deep-weathered materials. The concentrations of these ions increase from <10 to >40 m-equiv.l^{-1} with distance downvalley, up to 80 km from the aquifer's intake zone around granite domes. At the same time, pH de-

Photo 1: *This breakaway east of Sandstone (fig.3) clearly shows the effects of sapping of the soft, kaolinised materials beneath the duricrusted surface, leading to rockfall and, in this instance, a natural arch. Subsurface water movements are considered to play an important role in producing these characteristic landforms.*

creases from around neutral to <4. Other changes include a decrease in sodium, potassium, calcium, magnesium and sulphate relative to chloride, with distance along the aquifer and with increasing salinity. BETTENAY et al. (1964, 207) stated that: 'whilst it is usual to postulate that acidity in groundwater is due to weathering of pyrites, with consequent production of sulphuric acid, the evidence here appears to be in favour of an exchange mechanism involving the acid (aluminium) clays of the aquifer and the highly saline ground-waters.' Their data therefore suggest that deep weathering is a continuing process. This is further supported by the author's field observations of chemically-altering biotite and other mineral grains near the base of deep-weathered profiles further to the west, at Red Hill (fig.3).

3.4 Measurements of Eolian and Related Processes

The little research undertaken on eolian processes has been in relation to problems of accelerated erosion on farms (MARSH & CARTER 1983). No process measurements have been conducted on the saline playas, although KILLIGREW (1971), employing mineralogical and sedimentary analytical methods, showed that contemporary fluvial processes in uncleared, playa-bordering catchments are contributing sediments to some playa surfaces south of Kalgoorlie. GLASSFORD (1973), using scan-

Photo 2: *A lunette on the eastern edge of the saline playa, Lake Hurlstone, 200 km south of Southern Cross (fig.3). The photograph shows both the trapping of eolian, lake-derived sediments by the sparse, salt-tolerant vegetation colonising the lunette, and the effects of overland flow washing lunette materials back to the playa surface. The contemporary geomorphic result is a nice balance between eolian deposition and fluvial erosion.*

ning electron microscopy amongst other sedimentary analytical techniques (sediment bedform, bedding, particle size, particle shape, particle surface texture and mineralogy), demonstrated that one playa system near Hines Hill (and probably most others) is subjected to several contemporary processes: removal and redeposition of materials by wind and surface water (from both upvalley and upslope); expansion and contraction of the fine sediments by alternate wetting and drying, and intensive saline weathering **in situ** (GLASSFORD & CONACHER 1973). BOWLER's (1976) work indicates that the playa-fringing lunettes are paleofeatures; but direct observations suggest that deposition and removal of materials by wind are still continuing on some lunettes (photo 2), as is removal of materials by overland flow.

Process measurements have not been conducted on the sandplains (fig.2B). However, observations suggest that even where the indigenous vegetation is (briefly) destroyed by wildfire, little sand movement occurs on the exposed surface (photo 3). In other words the sandplains appear to be paleo-sediments. However, their origins are the subject of dispute.

From detailed sediment analyses over a regional transect from the coastal plain to the eastern edge of the Yilgarn block, KILLIGREW & GLASSFORD (1976),

Photo 3: *Despite the destruction by fire of this otherwise undisturbed heath vegetation, there were no surface ripple marks or other evidence of sand mobilisation by wind; suggesting a significant degree of stability of this sandplain unless physically disturbed by trampling or cultivation. The site is located east of the agricultural areas, some 220 km east of Belka (fig.3).*

GLASSFORD & KILLIGREW (1976) and especially GLASSFORD (1980) interpreted these coarse yellow sands as being wind-deposited, from an easterly direction, during several periods over a long geological history. GLASSFORD (1980) considered that there have been at least twenty 'major arid periods' and three major warm wet periods, relative to the present, over the past two million years. He stated that the arid periods were associated with global ice ages and, in southwestern Australia, cold, dry and strong winds, the suppression of rainfall from tropical cyclones, and eolian sand movements. The eolian sediments were classified by GLASSFORD (1980) into three groups:

1. Late Pliocene to Middle Pleistocene (ca. 4 million to 710,000 years BP), when both the calcretised and non-calcretised red sandy clay valley fills were deposited. The area of deposition was considered to have probably extended to the coastal zone;

2. Middle Pleistocene (710,000 to 128,000 years BP), when the most extensive deposition of yellow quartz sands occurred on the Yilgarn block and in the Perth Basin. GLASSFORD noted that these sands are dichronous and were probably extensively remobilised in the Late Pleistocene in <300 mm rainfall areas, and locally elsewhere; and

3. Late Pleistocene (128,000 to 10,000 years BP), during which the longitudinal desert dunes were deposited together with or followed by source-bordering lunate dunes (lunettes).

WYRWOLL & KING (1984) have rejected several of the above interpretations. They compared sand grain size distribution, rounding and surface textures from inland dunes, one sandplain site and coastal sands near Geraldton and Perth. The grain characteristics from the three location types were found to be significantly different. WYRWOLL & KING concluded that both the sandplain materials and the coastal sands are primarily residues from weathering—the former from a deep-weathered profile with only minor subsequent transport, and the latter an **in situ**, decalcified weathering product of coastal eolianites. Unfortunately, WYRWOLL & KING (1984) considered only the brief reports by GLASSFORD & KILLIGREW (1976) and KILLIGREW & GLASSFORD (1976) and did not address the data presented in GLASSFORD (1980). However, they (p.286) accept other evidence for previous 'periods of aridity in extreme south-western Australia during both glacial and interglacial stages'.

4 Problems with Process Measurements

There are numerous problems associated with process measurements and in relating the results of such measurements to landform evolution in this environment. Perhaps the two most crucial difficulties are the relationship of contemporary process rates to past processes and events, given the geological and climatic history, and the problem of scale and hence sampling strategies. These temporal and spatial problems are intimately related. Other difficulties include spatial heterogeneity, experimental design and operator error, and the effects of human impacts on geomorphic process/response systems.

4.1 Temporal and Spatial Problems

The marked seasonality of the climate gives rise to some distinctive problems. In winter, under natural vegetation, infiltration rates are high and overland flow rates are consequently very low. PUVANESWARAN (1981) found that overland flow constituted only 0 to 0.08% (and stemflow up to 4.7%) of individual rainfall events. In the drier inland areas, PILGRIM (1981) measured significantly higher proportions of overland flow (1.7–4.7%; calculated from his tables 5.1, 5.2, 5.3 and 5.12), but these are still low. In contrast, surface soil materials in summer are dessicated and often water-repellent due to an organic coating on individual grains (ROBERTS & CARBON 1972, McGHIE & POSNER 1980). Thus if the first winter rainfall is reasonably intense and of short duration, not permitting soil wetting to take place, infiltration rates are negligible. Virtually all the rain falling under such circumstances becomes overland flow with consequently high rates of erosion (McGHIE 1980). Compounding the problem for the researcher, this water repellency is not spatially uniform. Not only is it associated more with certain soil types than others, but the property also varies considerably over very short distances (ROBERTS & CARBON 1971).

Rain-bearing depressions, comprising

the remnants of tropical cyclones, bring intense rain over extensive areas in some summer seasons. For the reasons outlined above, the geomorphic significance of such events may be considerable. For example, during the eight-day period 17–24 February 1975, Cyclone Trixie produced falls almost double the annual average over a wide area centred near Sandstone (fig.3). A month later the resulting floodwaters washed away the Trans-Australia railway east of Zanthus, some 550 km to the south (SOUTHERN 1979). In any one area, however, the return period for such an event is unknown, but is likely to be at least ten years. For example in the ten years 1967–77, SOUTHERN (1979) plotted seven 'major tropical cyclones' which crossed into the southern half of Western Australia. Only two of the cyclones followed similar tracks.

Very intensive rainfalls associated with convective thunderstorms are much more localised and pose similar, but more severe, difficulties. The geomorphic work done is highly significant in comparison with that done by low-intensity, winter rains. In other words this is a classic magnitude/frequency problem. To obtain data amenable to time series analysis, in order to identify the natural erosion rates associated with rainfalls of various intensities and return periods, would require field-based monitoring over periods of several hundreds of years. Indeed, SOUTHERN (1979, 213) states that, in inland areas, 'the probability that a storm of maximum efficiency has been observed is low'.

Analyses of storm frequencies, magnitudes and durations have been undertaken for the Perth metropolitan coast by STEEDMAN AND ASSOCIATES (1982) and PANIZZA (1983). However, the storm data analysed extend back only to 1962 and 1968, respectively, although sufficient to indicate significant differences between seasons in different years. No similar analyses have been carried out for inland locations. Both researchers found a highly irregular temporal pattern of storminess which, whilst related to seasonal differences, is clearly unpredictable from one year to another. The difficulties this raises for the experimental field geomorphologist are obvious.

If process measurements are being undertaken to elucidate landform evolution, then the measurements need to be made of all processes and on all landform elements. PILGRIM (1981) and PUVANESWARAN (1981) focussed their overland flow measurements on landsurface unit 5 of the nine unit landsurface model on the assumption that this is the most rapidly changing element of the landsurface. But this assumption needs to be tested. This has only partly been done by PILGRIM in relation to landsurface unit 6, and by VALENTINE (1976), in a different context, in relation to unit 1. Moreover the ways in which changes of unit 5 affect overall landsurface geometry have yet to be investigated.

All the above assumes that contemporary processes are geomorphically significant. But 'process-response remnants' occur only in particular places (e.g. photo 1), and contemporary measurements of the rates of removal or alteration of the sediments or paleosols can only be done where the materials are present. Clearly, the remnants may not be representative of previous, more extensive deposits.

Similar difficulties occur with the need to find relatively undisturbed sites. Aboriginal occupation of much of this region probably extended over at least 40,000 years. Although they did not graze domestic animals or practise any form of cultivation, the Aborigines made extensive use of fire for a variety of purposes (HALLAM 1975). Thus the apparently-natural vegetation is not unaffected by human actions and may have been profoundly altered. A major transformation took place as a result of settlement by Europeans from 1829 onwards—particularly through the grazing of introduced stock on so-called natural 'pastures' (often woodlands), the clearing of the indigenous vegetation and its replacement with exotic, agricultural species, and the physical disturbance of the soils by cultivation. Remnant indigenous vegetation can be found in the southwest; but apart from the 1.8 m ha of State forests in the extreme west and southwest (fig.1), much of the vegetation retained in the agricultural areas is unrepresentative of the original cover. The areas left uncleared are usually those regarded as being unsuitable for agriculture - duricrusted ridges, rocky outcrops, areas with steep slopes or shallow soils, low-lying, swampy areas, and strips of land along some stream and river courses particularly where prone to flooding. As a consequence, process measurements within such remnants are very likely to produce results that are not applicable to the cleared areas.

The contribution of contemporary process measurements to the explanation of past landform development is much more problematic, even though, in most cases, the past processes may have been similar to present processes but operating with a greater or lesser intensity and/or frequency. This does not alter the fact that there are fundamental and possibly insurmountable difficulties involved in identifying and quantifying such modifications. Further, the action of past processes as is the case for contemporary processes, was certainly not spatially uniform. For assumed paleofeatures, such as relict duricrusts, sandplains, saline playas and lunettes, geomorphologists can perhaps look elsewhere for environments in which the relevant processes are currently taking place, but such analogue environments, if they exist, are unlikely to be exact replicas of the past southwestern Australian environments.

4.2 Problems of Scale

Overland flow measurements are carried out on small plots located on valley-sides or hillslopes. Extending the results of such measurements to total landsurfaces poses major difficulties—not only with respect to soil and slope heterogeneity, but also to regional climatic and lithological variations. Even with the use of a process/response spatial sampling framework such as the nine/unit landsurface model, the number of sites required to provide accurate data for the entire region on varying rates and forms of slope changes is enormous. In combination with the duration of the field monitoring needed, the provision of personnel, instrumentation and funds for such research becomes prohibitive.

Many streamflow discharges and subsurface water movements are measured in small catchments. The question of the extent to which the results of small catchment monitoring can be extended to larger drainage basins is still unresolved, although it has been considered by several researchers elsewhere (MORGAN 1974, KLEIN 1976, BOYD

1978). The indications are that, in general, rates of sediment movement per unit area from small catchments exceed the rates from larger basins (cf. STEERING COMMITTEE 1980). Even so, the spatial heterogeneity of slopes and soils within catchments and the palimpsest of paleo features would seem to render such simple relationships meaningless. The relationship of stream sediment discharge to slope sediment discharge is also unclear; but from work done elsewhere it appears to vary considerably both spatially (DUNNE & BLACK 1970, CAMPBELL 1977, BEVEN & KIRKBY 1979) and through time (GRAF 1983). Thus, again, results of stream sediment discharge measurements cannot, with present knowledge, be extrapolated to catchment denudation rates and especially not to the forms of slope changes. Moreover, the southwestern Australian stream gauging network does not routinely measure sediment discharge with the exception of total soluble salts; and there are very few gauging stations located in the drier areas.

5 Discussion and Conclusions

Process measurements have provided some information on the nature and contemporary rates of change of the southwestern Australian landsurface. However, consideration of the number of measurements required, spatially, and the period over which the measurements ideally need to be made, indicates that such data will at best be incomplete. It is impractical to plan for process measurements to be conducted at thousands of sites for hundreds of years. The most promising area for future work on contemporary processes appears to lie in the relationships between stream sediment discharges and slope erosion rates; and the most useful development for process geomorphology in this region would be the addition of sediment measurements to the existing stream monitoring network. To this needs to be added a carefully designed array of slope plots in both undisturbed and cleared, gauged catchments. If nothing else, an improved understanding of contemporary rates of geomorphic change in the landscape would provide benchmarks against which the effects of human impacts could be assessed.

It seems unlikely that process measurements can contribute a great deal to understanding how the southwestern Australian landsurface has changed in the past. A careful blending of process geomorphology with palynological, dendrochronological and archaeological research may provide important clues; but given the time spans involved, only the analysis of sediments is likely to yield relevant information. GLASSFORD (1980) and JOHNSTONE et al. (1973) have provided good examples of the application of sedimentary analysis to geomorphic interpretations. But relating an incomplete sedimentary record to processes and hence to changes in landsurface morphology through time is a tall order. It is in this context, perhaps, that laboratory experimentation and mathematical modelling can make significant contributions.

A number of geomorphologists are becoming increasingly concerned by the pursuit of apparent trivia by some of their process-orientated colleagues. It is argued that the original intention of process research—to explain landforms—has been forgotten. The underlying rationale of this paper has been to show that this admirable objective encounters some massive problems. Whilst the discussion has been in the context of the distinctive southwestern Australian environment, the difficulties are by no means unique to this region.

Acknowledgement

Dr. Ian Eliot of the Department of Geography, University of Western Australia, Dr. Adrian Harvey, Department of Geography, University of Liverpool, and an anonymous referee, made several constructive suggestions for which I am grateful; the remaining imperfections are the author's responsibility. The diagrams were drawn by Mrs. Michelle Bekle.

References

AUSTRALIAN BUREAU OF STATISTICS WESTERN AUSTRALIAN OFFICE (1977): Western Australian Year Book, Government Printer, Perth.

BATINI, F.E. & SELKIRK, A.B. (1978): Salinity sampling in the Helena catchment, Western Australia. Forests Department of Western Australia Research Paper **45**, Perth.

BETTENAY, E. (1962): The salt lake systems and their associated aeolian features in the semi-arid regions of Western Australia. Journal of Soil Science **13**, 10–17.

BETTENAY, E., BLACKMORE, A.V. & HINGSTON, F.J. (1964): Aspects of the hydrologic cycle and related salinity in the Belka valley, Western Australia. Australian Journal of Soil Research **2**, 187–210.

BETTENAY, E. & MULCAHY, M.J. (1972): Soil and landscape studies in Western Australia; 2. Valley form and surface features of the south-west drainage division. Journal of the Geological Society of Australia **8**, 359–369.

BEVEN, K. & KIRKBY, M.J. (1979): A physically based variable contributing area model of basin hydrology. Hydrological Sciences Bulletin **24**, 43–69.

BOWLER, J.M. (1976): Aridity in Australia: age, origins and expression in aeolian landforms and sediments. Earth Science Reviews **12**, 279–310.

BOYD, M.J. (1978): A storage-routing model relating drainage basin hydrology and geomorphology. Water Resources Research **14**, 921–928.

BREWER, R. & BETTENAY, E. (1973): Further evidence concerning the origin of the Western Australian sand plains. Journal of the Geological Society of Australia **19**, 533–541.

BROWN, D.A., CAMPBELL, K.S.W. & CROOK, K.A.W. (Eds.) (1968): The Geological Evolution of Australia and New Zealand. Pergamon, Oxford.

CAMPBELL, I.A. (1977): Sediment origin and sediment load in a semi-arid drainage basin. In: D.O. DOEHRING (Ed.), Geomorphology in Arid Regions, Allen and Unwin, London, 165–185.

CLEAVES, E.T., GODFREY, A.E. & BRICKER, O.P. (1970): Geochemical balance of a small watershed and its geomorphic implications. Bulletin of the Geological Society of America **81**, 3015–3032.

CONACHER, A.J. (1975): Throughflow as a mechanism responsible for excessive salinisation in non-irrigated, previously arable land in the Western Australian wheatbelt: a field study. CATENA **2**, 31–68.

CONACHER, A.J. (1982): Salt scalds and subsurface water: a special case of badland development in southwestern Australia. In: R. BRYAN & A. YAIR (Eds.), Badland Geomorphology and Piping, Geo Books, Norwich, 195–219.

CONACHER, A.J. & DALRYMPLE, J.B. (1977): The nine unit landsurface model: an approach to pedogeomorphic research. Geoderma **18**, 1–154.

CONACHER, A.J., NEVILLE, S.D. & KING, P.D. (1983): Evaluation of throughflow interceptors for controlling secondary soil and water salinity in dryland agricultural areas of southwestern Australia: II. Hydrological study. Applied Geography **7**, 115–132.

DUNNE, T. & BLACK, R.D. (1970): Partial area contributions to storm runoff in a small New England watershed. Water Resources Research **6**, 1296–1311.

FORESTS DEPARTMENT (1972): 1:500,000 map, Forest Areas of the south west, Perth.

GLASSFORD, D.K. (1973): A preliminary description and evaluation of surficial ephemeral fluvial sediment of the Yilgarn river, southwestern Australia. Unpub. B.A. (Hons.) Thesis in Geography, University of Western Australia, Perth.

GLASSFORD, D.K. (1980): Late Cenozoic desert eolian sedimentation in Western Australia. Unpub. PhD Thesis, University of Western Australia, Perth.

GLASSFORD, D.K. & CONACHER, A.J. (1973): A preliminary examination of sedimentary environments in an ephemeral drainage system. Paper presented to the 45th Australian and New Zealand Association for the Advancement of Science Congress, Perth.

GLASSFORD, D.K. & KILLIGREW, L.P. (1976): Evidence for Quaternary westward expansion of the Australian desert in southwestern Australia. Search **7**, 394–396.

GRAF, W.L. (1983): Variability of sediment removal in a semi-arid watershed. Water Resources Research **19**, 643–652.

HALLAM, S.J. (1975): Fire and Hearth: a Study of Aboriginal Usage and European Usurpation in South-Western Australia. Australian Institute of Aboriginal Studies, Canberra.

HAND, W.M. (1974): The relationship of clearing and rainfall to salinity of streams within the Manjimup woodchip licence area. Unpub. B.A. (Hons.) Thesis in Geography, University of Western Australia, Perth.

HINGSTON, F.J. & GALAITIS, V. (1976): The geographical variation of salts precipitated over Western Australia. Australian Journal of Soil Research **14**, 319–335.

JOHNSTONE, M.H., LOWRY, D.C. & QUILTY, P.G. (1973): The geology of southwestern Australia: a review. Journal of the Royal Society of Western Australia **56**, 5–15.

JUTSON, J.T. (1934): The Physiography (Geomorphology) of Western Australia. Geological Survey of Western Australia Bulletin **95** (revised edition of Bulletin **61**).

KILLIGREW, L.P. (1971): The identification of input sediments on several West Australian playas. Unpub. B.A. (Hons.) thesis in Geography, University of Western Australia, Perth.

KILLIGREW, L.P. & GLASSFORD, D.K. (1976): Origin and significance of oolitic kaolin spherites in sediments of south-western Australia. Search **7**, 393–394.

KLEIN, M. (1976): The influence of drainage area in producing thresholds for the hydrological regime and channel characteristics of natural rivers. University of Leeds Department of Geography Working Paper **147**.

MARSH, B. & CARTER, D. (1983): Wind erosion. Western Australian Journal of Agriculture **24**, 54–57.

McARTHUR, W.M. & BETTENAY, E. (1979): The land. In: B.J. O'BRIEN (Ed.), Environment and Science, University of Western Australia Press, Perth, 22–52.

McGHIE, D.A. (1980): The contribution of the mallet hill surface to runoff and erosion in the Narrogin region of Western Australia. Australian Journal of Soil Research **18**, 299–307.

McGHIE, D.A. & POSNER, A.M. (1980): Water repellence of a heavy-textured Western Australian surface soil. Australian Journal of Soil Research **18**, 209–223.

MORGAN, R.P.C. (1974): Problems of scale affecting the extrapolation of small catchment studies to larger basins. Paper presented to British Geomorphological Research Group Basin Sediment Systems Sub-Group Meeting, London School of Economics, London.

MULCAHY, M.J. (1971): Landscapes, laterites and soils in south Western Australia. In: J.N. JENNINGS & J.A. MABBUTT (Eds.), Landform Studies from Australia and New Guinea, Australian National University Press, Canberra, 211–230.

MULCAHY, M.J. (1973): Landforms and soils of southwestern Australia. Journal of the Royal Society of Western Australia **56**, 16–22.

PANIZZA, V. (1983): Westerly storms of the Perth metropolitan coast, Western Australia. Unpub. BSc (Hons) Thesis in Geography, University of Western Australia, Perth.

PECK, A.J. (1978): Salinisation of non-irrigated soils and associated streams: a review. Australian Journal of Soil Research **16**, 157–168.

PECK, A.J. & HURLE, D.H. (1973): Chloride balance of some farmed and forested catchments in southwestern Australia. Water Resources Research **9**, 648–657.

PECK, A.J., THOMAS, J.F. & WILLIAMSON, D.R. (1983): Effects of man on salinity in Australia. Water 2000 Consultant Report **8**, Department of National Development and Energy, Canberra.

PILGRIM, A.T. (1979): Landforms. In: J. GENTILLI (Ed.), Western Landscapes, University of Western Australia Press, Perth, 49–87.

PILGRIM, A.T. (1981): Spatial variability of hydrologic response on naturally vegetated hillslopes in a semi-arid environment. Unpub. PhD thesis, University of Oklahoma, Norman.

PILGRIM, A.T. & CONACHER, A.J. (1974): Causes of earthflows in the southern Chittering valley, Western Australia. Australian Geographical Studies **12**, 38–56.

PILGRIM, A.T., PUVANESWARAN, P. & CONACHER, A.J. (1986): Factors affecting natural rates of slope development. CATENA **13**, 169–180.

PRIDER, R.T. (1977): Physical features and geology. In: Western Australian Year Book, Government Printer, Perth, 19–38.

PUVANESWARAN, P. (1981): Soil-slope relationships in the Wungong Brook catchment, Western Australia. Unpub. M.A. Thesis in Geography, University of Western Australia, Perth.

PUVANESWARAN, P. & CONACHER, A.J. (1983): Extrapolation of short-term process data to long-term landform development: a case study from southwestern Australia. CATENA **10**, 321–337.

ROBERTS, F.J. & CARBON, B.A. (1971): Water repellence in sandy soils of south-western Australia. I. Some studies related to field occurrence. Division of Plant Industries, Commonwealth Scientific and Industrial Research Organisation Field Station Records **10**, 13–20.

ROBERTS, F.J. & CARBON, B.A. (1972): Water repellence in sandy soils of south-western Australia. II. Some chemical characteristics of the hydrophobic skins. Australian Journal of Soil Research **10**, 35–42.

SHEA, S.R. & HATCH, A.B. (1976): Stream and groundwater salinity levels in the South Dandalup catchment of Western Australia. Forests Department of Western Australia Research Paper **22**, Perth.

SHERWOOD, M.J. (1969): An examination of the source of a stream in a semi-arid enviroment. Unpub. M.A. (Prelim.) Thesis in Geography, University of Western Australia, Perth.

SOUTHERN, R.L. (1979): The atmosphere. In: B.J. O'BRIEN (Ed.), Environment and Science, University of Western Australia Press, Perth, 183–226.

STEEDMAN, R.K. & ASSOCIATES (1982): Record of Storms, Port of Fremantle 1962–1980. Public Works Department, Perth.

STEERING COMMITTEE (1978): Report of the Steering Committee on research into the Effects of Bauxite Mining on the Water Resources of the Darling Range, Western Australian Department of Industrial Development, Perth.

STEERING COMMITTEE (1980): Report by the Steering Committee on Research into the Effects of the Woodchip Industry on Water Resources in South Western Australia. Department of Conservation and Environment Bulletin **81**, Perth.

VALENTINE, P.S. (1976): A preliminary investigation into the effects of clear cutting and burning on selected soil properties in the Pemberton area of Western Australia. University of Western Australia Department of Geography Geowest **8**, Perth.

WILLIAMSON, D.R. & BETTENAY, E. (1979): Agricultural land use and its effect on catchment output of salt and water - evidence from southern Australia. Progress in Water Technology **11**, 463–480.

WYRWOLL, K.-H. (1979): Late Quaternary climates of Western Australia: evidence and mechanism. Journal of the Royal Society of Western Australia **62**, 129–142.

WYRWOLL, K.-H. & KING, P.D. (1984): A criticism of the proposed regional extent of Late Cenozoic arid zone advances into south-western Australia. CATENA **11**, 273–288.

WYRWOLL, K.-H. & MILTON, D. (1976): Widespread Late Quaternary aridity in Western Australia. Nature **264**, 429–430.

Address of author:
A.J. Conacher
Department of Geography
University of Western Australia
Nedlands WA 6009
Australia

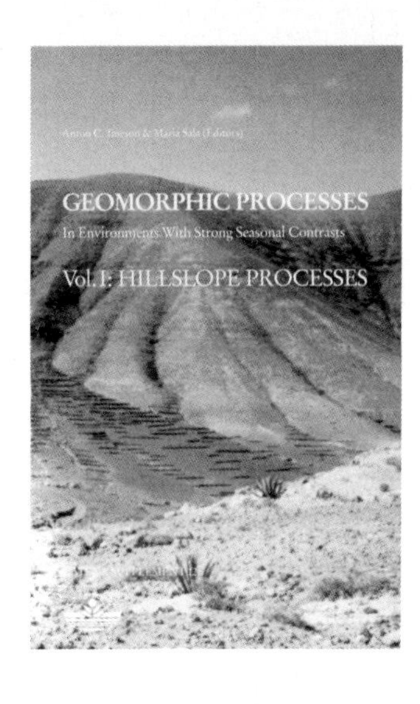

Anton C. Imeson & Maria Sala:

GEOMORPHIC PROCESSES

In Environments With Strong
Seasonal Contrasts
Vol. I: HILLSLOPE PROCESSES

CATENA SUPPLEMENT 12, 1988

Price: DM 149,— / US $88.—

ISSN 0722-0723 / ISBN 3-923881-12-3

CONTENTS

N E W

CATENA paperback

Joerg Richter

THE SOIL AS A REACTOR

Modelling Processes in the Soil

If we are to solve the pressing economic and ecological problems in agriculture, horticulture and forestry, and also with "waste" land and industrial emmissions, we must understand the processes that are going on in the soil. Ideally, we should be able to treat these processes quantitatively, using the same methods the civil engineer needs to get the optimum yield out of his plant. However, it seems very questionable, whether we would use our soils properly by trying to obtain the highest profit through maximum yield. It is vital to remember that soils are vulnerable or even destructible although or even because our western industrialized agriculture produces much more food on a smaller area than some ten years ago.

This book is primarily oriented on methodology. Starting with the phenomena of the different components of the soils, it describes their physical parameter functions and the mathematical models for transport and transformation processes in the soil. To treat the processes operationally, simple simulation models for practical applications are included in each chapter.

After dealing in the principal sections of each chapter with heat conduction and the soil regimes of material components like gases, water and ions, simple models of the behaviour of nutrients, herbicides and heavy metals in the soil are presented. These show how modelling may help to solve problems of environmental protection. In the concluding chapter, the problem of modelling salt transport in heterogeneous soils is discussed.

The book is intended for all scientists and students who are interested in applied soil science, especially in using soils effectively and carefully for growing plants: applied pedologists, land reclamation and improvement specialists, ecologists and environmentalists, agriculturalists, horticulturists, foresters, biologists (especially microbiologists), landscape planers and all kinds of geoscientists.

Prof.Dr. Joerg Richter
Institute of Soil Science
University of Hannover, FRG

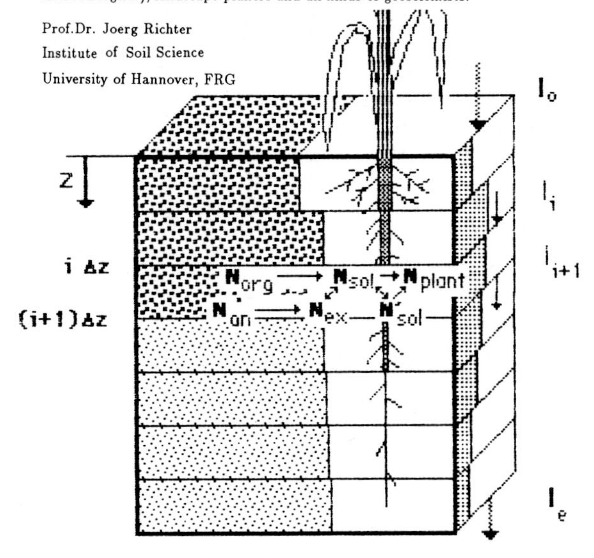

ISBN 3-923381-09-3

Price: DM 38,50 / US $ 24.—